Vinícius Lurentt Bourguignon

Bioethics in science education

Vinícius Lurentt Bourguignon

Bioethics in science education

The importance of empathy and insight

ScienciaScripts

Imprint

Any brand names and product names mentioned in this book are subject to trademark, brand or patent protection and are trademarks or registered trademarks of their respective holders. The use of brand names, product names, common names, trade names, product descriptions etc. even without a particular marking in this work is in no way to be construed to mean that such names may be regarded as unrestricted in respect of trademark and brand protection legislation and could thus be used by anyone.

Cover image: www.ingimage.com

This book is a translation from the original published under ISBN 978-613-9-60597-2.

Publisher:
Sciencia Scripts
is a trademark of
Dodo Books Indian Ocean Ltd. and OmniScriptum S.R.L publishing group

120 High Road, East Finchley, London, N2 9ED, United Kingdom
Str. Armeneasca 28/1, office 1, Chisinau MD-2012, Republic of Moldova, Europe
Printed at: see last page
ISBN: 978-620-6-01749-3

Summary

PRESENTATION

The article published in this book, was the result of a research conducted during the author's graduation at the University Vila Velha in the year 2013 as a conclusion work of the Biological Sciences course.

While attending the biological sciences college I had little or no contact with the study of Newtonian psychology until I met important authors of constructivist pedagogy such as Piaget and Vygotsky during the degree course in Nature Sciences at the Instituto Federal Fluminense (Campos Centro) in the year 2016 in my brief passage through the institution. Therefore, I would like to present here a summary of *Chapter 6 - The Newtonian psychology of* the book *The Turning Point* by Fritjjof Capra and then present the main ideas of Lev Vygotsky on the social formation of the mind.

Newtonian psychology

In the words of Fritjjof Capra[1] , "psychology would have emerged following the Cartesian model of Descartes, however the idea of a mind separated from a body made it very difficult to understand the psychological phenomena, and this Cartesian model gave rise to two schools of thought in the study of psychology, structuralism and behaviorism; the structuralists sought to understand the mind through introspection and analysed consciousness in its most basic elements, while the behaviourists understood the functioning of the mind from behaviour, leading them to ignore or deny the pure and simple existence of the mind; both currents emerged in a period dominated by the Newtonian model of reality, incorporating basic concepts of mechanics into their basic structure and it was then that Freud developed psychoanalysis by the method of free association." - According to Capra - "it is believed that psychology as a science would have emerged in the 19th century and its historical roots would be attributed to ancient Greece, however the Western belief that this tradition was the only one to produce serious psychological theories is now being recognised as a narrow and culturally conditioned view". - For Capra - "recent achievements in research on psychology have stimulated interest in Eastern systems of thought, especially those of India, for contemplating both the materialistic and idealistic views, from monism to dualism, with numerous theories" - so for him - "in addition to the philosophical wealth, Eastern cultures have developed spiritual traditions that are based on empirical knowledge, approaching the approach of modern science, and " - so, for Capra - "these traditions although not following the Cartesian model, have led to create elaborate models of consciousness that are in agreement with recent scientific achievements." - So he continues. - "During their history, they have developed subtle techniques to change in their adherents the conscious perception of their own existence and their relationship with society and the natural world, and traditions such as yoga and Taoism resemble psychotherapies much more than philosophies or religions and the psychological speculations of the ancient Greek philosophers also reveal strong influences of Eastern ideas and this early Western psychology diverges between idealistic and materialistic conceptions of the soul. Empedocles taught a materialist theory of the psyche, according to which thought and perceptions depended on bodily changes. Pythagoras on the other hand expounded mystical conceptions such as the transmigration of souls. Socrates used the word 'psyche' in the sense that is given by modern psychology, as the seat of intelligence and character, in place of the mystical concept described earlier as vital force. Plato was the first to deal explicitly with the problem of consciousness, and Aristotle wrote the first treatise in this regard, in which he developed a biological and materialistic approach to psychology". - On the other hand, Capra states - "this materialist approach, subsequently elaborated by the Stoics, found resistance in Plotinus and Neoplatonism, whose teachings were based on the Indian Vedanta philosophy that would have a strong influence on early Christianity, where according to Plotinus the soul is immaterial and immortal, the consciousness is the image of divinity and as such is present at all levels of reality. In Fedro, Plato describes the soul as an auriga leading two horses, one representing the passions of the body and the other the higher emotions. This conflict generated the "mind-body" problem that was reflected in many schools of psychology, especially between Freud and Jung." - Moreover, continues Capra - "in the 17th century the "mind-body" problem shaped the subsequent development of Western scientific psychology, and for Descartes, mind and body were parallel domains, and could be studied independently of each other, - thus - the body was governed by mechanical laws, but the mind was free and immortal - for him - the soul or consciousness could affect the body by interacting through the pineal gland of the brain and human emotions were described in a semi-mechanical way. - For Descartes, in Capra's interpretation - "knowledge was a primary function of human reason, taking place independently of the brain and the concepts that played such an important role in Descartes' philosophy and science could not be derived from the confused performance of the senses, but was the result of an innate cognitive disposition."

"Inspired by the Cartesian paradigm, two great philosophers of the 17th century, Spinoza and Leibniz introduced their own ideas. Spinoza contrary to Descartes' dualism replaced it by monism, something mystical, while Leibniz introduced the existence of nomads, essentially psychic units of the organism that reflected each other. There was therefore no interaction between body and mind, but both acted in "pre-established harmony". The development of psychology did not follow the views of Spinoza or Leibniz. Instead philosophers turned to Newton's precise mathematical formulation of Descartes' mechanistic paradigm and tried to use its principles to understand human nature. While La Mettrie in France applied the mechanical model of animals to humans, including their minds, British empiricist philosophers used Newtonian ideas to develop more refined psychological theories. Hobbes and Lock refuted the Cartesian concept of innate ideas and held that nothing existed in the mind that had not passed through the senses. For Lock, the human mind was at its birth a tabula rasa, on which ideas were recorded through sensory perceptions. From there a mechanistic theory of knowledge could be formulated, in which emotions were the basic elements of the mental domain, combining into more complex structures by the process of association. David Hume elevated association to a central principle in the analysis of the human mind, seeing it as an "attraction in the mental world", which was akin to gravitational force in the Newtonian material universe. And influenced by Newton's method of inductive reasoning he reduced the "self" to a "bundle of perceptions". David Hartley went a step further by developing a detailed and ingenious mechanistic model of the mind in which all mental activity was reduced to neurophysiological processes. This model was further elaborated in the 1870s by Wilhelm Wundt, considered the founder of scientific psychology. Modern psychology was the result of advances in anatomy and physiology in the 19th century. The mechanistic models described by Descartes, La Mattrie and Hartley were recast in modern terms, and the Newtonian orientation became firmly established in psychology."

"The discovery of correlations between mental activity and the structure of the brain led some neuroanatomists to postulate that human behavior could be reduced to a set of independent mental faculties or traits, located in specific regions of the brain, and at first researchers were able to demonstrate where the primary motor and sensory functions were located, but when the approach was extended to higher cognitive processes, such as learning and memory, no coherent picture of these phenomena was found. With 19th century studies of the nervous system came another field of research, reflexology; the relation between stimulus and response and its mechanical reliability, which formed the basis of all the more complex patterns of behavior. This theory was expounded by Ivan Sechenov, founder of the Russian school of reflexology, whose distinguished member was Ivan Pavlov. Pavlov's discovery of the principle of conditioned reflexes had a decisive impact on subsequent theories of learning. Detailed investigation of the central nervous system helped to establish systematic relations between the quality of sensory experiences and the physical characteristics of their stimuli. Pioneering experiments by Ernst Weber and Gustav Fechner resulted in the formulation of the Weber-Fechner law, which postulates a mathematical relationship between the intensities of sensations and their stimuli. Physicists, such as Hermann von Helmholtz, offered important contributions to this field of sensory physiology, for example in the development of comprehensive theories of colour vision and hearing. Influenced by studies of perception and behavior, Wundt was the main representative of the so-called elementalist orientation, which held that all mental functioning could be analyzed in specific elements. For Wundt, the object of psychology was the study of how these elements could combine to form perceptions, ideas and various associative processes".

"The orthodox experimental psychologists of the 19th century were dualists and saw introspection as an analytical method that would allow them to reduce consciousness to well-defined elements associated with specific nerve currents in the brain. These theories aroused strong opposition among psychologists who emphasized the unitary nature of consciousness and perception, and thus gave rise to two influential schools: gestaltism and functionalism". - For Capra - "both incapable of changing the Newtonian orientation followed by most psychologists during the following century" - and further - "strongly influenced the new trends in psychology and psychotherapy that emerged in the second half of our century and the gestalt psychology founded by Max Wertheimer and his collaborators was based on the assumption that 'living organisms do not perceive things in terms of isolated elements, but in terms of Gestalten', that is, 'meaningful totalities that exhibit qualities absent in each of their individual parts'." - According to Capra, - "Kurt Goldstein is said to have applied this method with the aim of helping people to harmonise with themselves and their environment. The development of functionalism was a consequence of 19th century evolutionary thinking, and the relationship between structure and function. For Darwin each anatomical structure was a functional component of an integrated living organism engaged in the evolutionary struggle for survival. Psychologists then shift from the study of mental structure to the study of mental processes and their functioning. These functionalist psychologists emphasize the unity and dynamic nature of the 'stream of consciousness'. The functionalist William James, considered by many to be the greatest American psychologist, was a vehement critic of atomistic and mechanistic tendencies in psychology and an advocate of the interdependent interplay of body and mind. He reinterpreted the experimental findings of his time, emphasizing consciousness as a personal, integral, and continuous phenomenon, and in 1890 published his innovative views on the human psyche in the monumental *Principles of Psychology*, which soon became a classic.

"In the 20th century psychology made great progress and acquired increasing prestige. It benefited considerably from cooperation with other disciplines and found important applications in health care, education and many other areas of human activity. Since then, psychological thought has been dominated by two powerful schools, behaviourism and psychoanalysis. Behaviourism represents the culmination of the mechanistic approach in psychology, they created a "soulless psychology", a refined version of La Mettrie's human machine. Mental phenomena were reduced to types of behaviour, and behaviour, to physiological processes governed by the laws of physics and chemistry. John Watson who founded behaviourism was strongly influenced by various trends in the human sciences at the beginning of the century."

"Edward Titcher, an acknowledged leader of the "structuralist" school, attempted a strict reduction of contents of consciousness to "simple" elements and stressed that the "meaning" of mental states was nothing more than the context in which mental structures were found - and for Capra - presenting no additional contribution to psychology. At the same time, the reductionist and materialist conception of mental phenomena was influenced by Loeb's mechanistic biology and his theory of tropism. Loeb explained the tendency of plants and animals to turn certain parts of themselves in certain directions in "forced movements" imposed on living organisms by the environment in a strictly mechanistic way. This new theory exerted an enormous attraction on many psychologists, who applied the notion of forced movements to a wider range of animal behaviors and eventually to those of human beings.[1]

After all these advances in psychology, it was Lev Vygotsky who was responsible for presenting a psychology of the mind socially constructed from a historical context, which only became known worldwide after its prohibition and rediscovery between 1956 and 1962, and in 1984 in Brazil when the book *A formação social da mente* was published. Besides being a Marxist, Vygotsky was influenced by the functionalist school and by Wolfgang Kohler's Gestalt psychology, elaborating a theory of mind much more connected to the "mental functions" and their interactions with matter, and from that to its social construction.[2]

Vygotsky and the social formation of the mind

Despite the significant advances that can be attributed to the behaviourist methodology, Vygotsky states that this method presents series limitations, and although the method is objective and suitable for the study of simple reflexes, it clearly fails when applied to complex psychological processes, because the innermost mechanisms characteristic of these processes remain hidden.[3]

The pedagogical teacher Teresa Cristina Rego[2] identified some central ideas about the social formation of the mind in *Vygotsky*. The first idea refers to the relation individual/society, and in the words of the pedagogue, "the typically human characteristics are not present from the individual's birth, nor are they merely the result of pressures from the external environment, they result from the *dialectical interaction of* man and his socio-cultural environment, at the same time that the human being transforms his environment to meet his basic needs, he transforms himself". In the words of Luria, quoted by Teresa Rego, "the higher psychological functions of the human being arise from the interaction of biological factors, which are part of the physical constitution of Homo sapiens, with cultural factors, which have evolved over tens of thousands of years of human history".

The second idea refers to the cultural origin - since, in Teresa Rego's interpretation - "the human mental development is not given a *priori, it* is not immutable and universal, it is not passive, nor independent of the historical development and the social forms of human life" - therefore, for Vygotsky, in this interpretation - "culture is part of human nature, since its psychological characteristics are given through the internalization of historically determined and culturally organized ways of operating information". The third idea refers to the biological basis of psychological functioning, the brain, which is understood as an "open system, of great plasticity, whose structure and modes of operation are shaped throughout the history of the species and individual development".

The fourth idea assumes that mediation is fundamental in the social historical perspective because it is precisely through tools and signs that the processes of psychological functioning are provided, highlighting the role of language in the thought process. The fifth idea is based on the fact that "complex psychological processes differ from the most elementary mechanisms and cannot be reduced to a chain of reflexes".[2]

Furthermore Vygotsky noted the importance of perception; perception is an important psychological function for better understanding complex psychological processes and these more elaborate types of psychological functionings develop in a dialectical process, and can be explained and described from the historical and social context.

The importance of perception

For Lev Vygotsky the relationship between the use of the instrument and speech affects various psychological functions in particular perception, sensorimotor operations and attention, each one composing the dynamic system of behaviour, and emphasizes the work of Kohler on the importance of the structure of the visual field in the organisation of the practical behaviour of anthropoid monkeys and the use of perception to solve problems. And for Kohler, the difference between the practical intelligence of children and that of animals is that the former are able to reconstruct their perception and thus free themselves from a certain perceptual field structure, and this development represents a fundamental break with the natural history of primitive animal behaviour for the higher intellectual activities of human beings.[3]

According to Arno Engelmann, Gestalt would not be an attribute to oppose to the elements, on the contrary, it would be the way according to which things would be perceived, composed of forming elements or sensations, however, for Kohler there may be sensations of which one is not conscious and that even the conscious ones may lead to illusions, showing clearly that a simple geometry of the elementary stimulations disposed externally is incapable of explaining the observed facts, while Koffka in *The Psychology of Perception,* will say that basically it is not the sensations which form the perception, but it is the perception which can be composed of sensations, and for Petermann the perceptible datum is something immediate. Moreover, Lipps in 1897 said that in empathy there is a consciousness of the I which projects itself into the thing and infers an emotion to it, while Kohler will say that simply in the way in which he can feel an emotion in it, he can also put it outside the I. In any case, for Engelmann, in distinguishing the self from what I perceive outside myself, we make the division between the "phenomenally subjective" and the "phenomenally objective", between what "is inside" and what "is outside". The centre for him is also the place where the "vectors" originate, as in physics, and this way of seeing things constitutes the so-called dynamic interpretation of the world.[4]

According to Engelmann, Kohler calls emotions *tertiary qualities,* in comparison with *secondary* and *primary qualities according to* the old classification of Locke, where *primary qualities* are inseparably in the bodies, like solidity and extension, and *secondary qualities* which produce sensations but are not in the bodies, like colours and tastes. According to Engelmann every scientist has a series of facts to explain and these facts should be totally neutral to him, however, according to him, for Kohler every value is at the same time a fact, and values are learned by one's phenomenology, this phenomenology is isomorphic to an electric field in the brain and this electric field in the brain is identical with the electric field in inanimate matter. Therefore for Kohler, in Engelmann's interpretation, a repulsion or an attraction could be produced in inanimate matter as well as in the brain or in the phenomenological field. For Engelmann ethics, which deals with values, is at the apex of Kohler's thinking, and can be found together with other sciences, and according to Kohler himself psychology is a branch of biology.[4]

THANKS

I thank first of all nature's manifestations of love, our source of inspiration for a better life. I have to thank my teachers and the advisor Rafael Alvez Rezende (UVV), for the support and freedom he gave me to write this work keeping my references and objectives. I also thank professor João Luiz Rossi Junior, coordinator of CEUA-UVV at the time, for the opportunity he gave me by giving me space and access to the projects to carry out the research, and also for his presence in the board of examiners. I must thank Professor Rosana Suemi

Tokumaru (UFES) for showing interest in the research, for accepting the invitation to participate in the board, and for her insightful observations. I am very grateful to the biostatistics professor Romildo Rocha Azevedo Junior (UVV) for his guidance in the statistical analyses, and to my colleague Gabriel Santos (UFES) for his valuable tips. I thank all my colleagues and former colleagues, for so many exchanges of experiences, and the professors and teachers both at UVV and IFF Campos Centro for their teachings in the classroom. I want to thank everyone who passed by and especially those who stayed with me in this journey of great learning, especially my mother, my family and friends. I must also remember to thank the publishers who accepted the publication of the article produced. I thank the editors of the Brazilian Journal of Animal Law, journal that published the article a few months after my graduation in 2014. My thanks to Novas Edições Acadêmicas for the invitation a few years later to publish the article in book format. I also thank Atena Editora for the proposal and publication of the article in the electronic journal Pensando as Licenciaturas 2. My sincere thanks to all involved.

CHAPTER 1

Introduction

Science has always been the target of criticism regarding ethics and the way researchers sustain their methods to assert their power over nature. The famous phrase "playing god" arose from the practices of researchers in creating and manipulating life. As if that were not enough, they still imprison living, conscious animals for painful experiments until death. According to the conservatives, in order to advance research they claim the need to use live animals as models. Creating conscious life and making it suffer because of our curiosity and thirst for power is still a reality in research centres. But contrary to what many people think, this ends up imposing a block in relation to the possibilities that science can produce, and science has been advancing a lot in these last decades, enabling the creation of alternative methods to study organisms with less invasive methods, or even with the substitution of organs, tissues and cells, or even by artificial models, simulation and computers. And thinking about this, it is necessary to reflect on the ethical limits of animal experimentation.

Ethics is the philosophical reflection on morality, that is, on the rules and moral codes that guide human conduct[5] . According to Marilena Chauí:

> "Socratic questions inaugurate ethics or moral philosophy, because they define the field in which moral values and obligations can be established, by finding their starting point: the conscience of the moral agent. Only he who knows what he is doing, knows the causes and ends of his action, the meaning of his intentions and attitudes and the essence of moral values, is a moral ethical subject."[6]

Ethics is a struggle against our passions and instincts, it starts from the reflective consciousness of the individual:

> "Moral conscience manifests itself, first of all, in the capacity to deliberate in the face of possible alternatives, deciding and choosing one of them before launching into action. It has the capacity to evaluate and weigh personal motivations, the demands made by the situation, the consequences for oneself and for others, the conformity between means and ends (employing immoral means to achieve moral ends is impossible), the obligation to respect the
>
> established or to transgress it (if the established is immoral or unjust).
> The will is that deliberative and decisional power of the moral agent. In order to exercise such power over the moral subject, the will must be free, that is, it cannot be subjected to the will of another nor can it be subjected to the instincts and passions, but, on the contrary, it

must have power over them and them."[6]

In the words of Sonia Felipe[7] , "all moral action depends on reasoning that takes into consideration the unity and coherence of one's own acts in the face of the recognition of duty (rational desire) to practice them, and in this sense ethics, as expressed by Singer interpreted by Robert C. Solomon, is nothing more than 'a process of expanded consciousness', 'the expanding circle'" - for him - "reason allows the expansion of the circle, the overcoming of the primitive impulse to care for the good of the offspring, to the sphere that surpasses the family, the village, the country, the nation, the species".[7] In his interpretation, Solomon gave priority to reason in the process of expanding the 'moral circle', however, Frans de Waal[8] notes the 'expanding circle' in non-human animals beyond the species by the capacity for empathy, without the need for the use of reason as we know it in humans. "Empathy is the capacity that mammals have to put themselves in the other's place and experience their emotions" - as stated by Theodor Lipps quoted by Frans de Waal - "indirectly we enter their body and participate in their experience".

For Tréz & Nakada (2008), "a scientific practice that directly affects the perceptions and actions of those working in the biomedical and biological sciences, characterising the moral *status* assigned to non-human animals, is the use of the "animal model" in experimentation, and an analysis of this practice can help visualise how the anthropocentric-speciesist paradigm permeates the study and practice of modern biology".[9] And due to the ethical conflicts existing on the part of teachers and students in the study of biological sciences[9,10] - it is necessary to take a critical look at the methods used in the teaching of science - because studies show that alternative methods can generate results as significant as methods that use animals.[11,12]

Several thinkers have tried to propose a break from this hegemonic paradigm. The Norwegian philosopher Arne Naess, proposed in 1973 the *Deep Ecology* as an alternative to the hegemonic model (Table 1) to think of man as the centre of nature. In Capra's interpretation (apud SIQUEIRA-BATISTA, 2009):

> "[...] deep ecology does not separate human beings or anything else from the natural environment. It sees the world not as a collection of isolated objects, but as a network of phenomena that are fundamentally interconnected and are interdependent. Deep ecology recognizes the intrinsic value of all living things and conceives of human beings only as a particular thread in the web of life." [13]

The *deep* ecology brings in its studies a change of paradigm of the ecological perspective, where in its centre are the human-nature interactions on the vision of ethics and bioethics, both related to the biotic elements[13] and the abiotic ones according to other authors.[14]

11

Table 1. Comparison between the hegemonic worldview and deep ecology.

HEGEMONIC WORLDVIEW	DEEP ECOLOGY
Nature	Harmony with Nature
Natural environment as a resource for human beings	All Nature has intrinsic value
Human beings are superior to other living beings	Equality between different species
Economic and material growth as a basis for human growth	Material goals at the service of greater goals of self-realisation
Belief in ample reserves of resources	Planet has limited resources
Progress and solutions based on high technology	Appropriate technology and non-dominant science
Consumerism	Doing what is necessary and recycling
Centralised national community	Bioregions and recognition of minority traditions

Source: GOLDIM JR, 2005 apud SIQUEIRA-BATISTA et al, 2009.

In relation to abiotic elements, according to Singer[14] the arguments are more difficult to sustain; "we will tread on safer ground if we confine ourselves to reasoning grounded in the interests of sentient creatures".

Some authors who collaborated against the hegemonic worldview were the anarchists Élisée Reclus and Piotr Kropotkin. Élisée Reclus, according to Leal[15] , made significant contributions in relation to bioethics of space, the recognition of bioregions, the preservation of the environment and minority traditions. Kropotkin[16] contributed in his work *Mutual Support: A Factor of Evolution* with publications translated since 1902, where the author, based on Darwinian theories, argues about an important factor for the survival of species: mutualism. Kropotkin disagreed with the view of some of Darwin's followers who attributed to competition and conflicts one of the main factors of natural selection. For Kropotkin, it is not competition, but avoiding or reducing conflicts which guarantees the best survival of populations, and he even noted the importance of mutual support and association between workers to accelerate the process of civilization development. Kropotkin demonstrated through observations of nature that the more social and - consequently - the more populous species had greater chances of surviving extinctions caused by environmental pressures. Kropotkin also came to attribute an instinctive morality to animals, similar to what Proudhon[17] observed.

That is why I consider here important the biosocial relationships for human development, because according to Bronfenbrenner quoted by Paola Biasoli Alves[18] , human development is defined as "the set of processes through which the particularities of the person and the environment interact to produce constancy and change in the characteristics of the person in the course of his life".

For Van Renselae Potter[19] humanity urgently needs a new wisdom that provides "the knowledge of how to use knowledge" for human survival and for the improvement of the quality of life, and suggests the term bioethics for this science of survival that according to him should be built on the science of biology and ethics. Here I consider knowledge according to David Hume quoted by Matos[20] , who says "every animal presents instinctively, knowledge being a network of information generated by habit" - in this interpretation - "the structures of knowledge in the human being, and those similar in other living beings, are explained by taking into account their development through natural processes, such as natural selection".

For Pegoraro[21] , bioethics, besides being an "applied ethics", is a "philosophical ethics that specializes in accompanying the progress and ethical problems of genetics, biomedicine, the biosphere and ecosystems". According to Diniz[22] , "bioethics is concerned with all life situations that are in the midst of different moral choices as to the standards of well-living".

According to Chaui, "when we follow the history of ethical ideas we can see that at its centre is the problem of violence and the means to avoid, reduce or control it, and different social and cultural formations have instituted sets of ethical values as standards of social behavior that could ensure the physical and psychological integrity of its members and the conservation of the social group.[6] Still according to her, the various cultures and societies have not and do not define violence in the same way, and yet certain aspects of violence are perceived in the same way in the various cultures and societies, forming the common background against which ethical values are erected. According to Chaui:

> "...violence is perceived as the exercise of physical force and psychological coercion to compel someone to do something contrary to himself, contrary to his interests and desires, contrary to his body and conscience, causing him deep and irreparable harm, such as death, madness, self-aggression or aggression towards others."[6]

For Chaui, "when a culture and a society define what they understand as evil, crime and vice, they determine what they judge as violence against an individual or against the group and, simultaneously, they erect the positive values, the good and virtue, as ethical barriers against violence" and furthermore, "our culture and society define us as subjects of knowledge and action, from an ethical point of view, we are people and cannot be treated as things, and for this, ethical values offer a guarantee of our condition as subjects, morally forbidding that we are transformed into a thing used and manipulated by others".[6]

The concept of person, as limited as it is for the human moral community, for philosophers of animal ethics this concept also applies to non-human animals. According to Napoli[23] , Regan uses the concept of subjects-of-a-life, while they are persons for Francione; "persons are sentient and free individuals, that is, they are nobody's property".

Still for Chauí, ethics is normative precisely because of this, its norms aim to impose limits and controls on the permanent risk of violence.[6]

CHAPTER 2

Contributions of biology and philosophy to moral consideration with non-human animals

According to professor and philosopher Dr. Sonia Felipe[24] the human mechanism of pain is practically the same as in animals, and we know from experience, with humans, that the ability to feel pain is essential for survival. The sensation of pain, and the motivational influence of feeling it, are essential to the survival of the system, and to suggest that the mechanism is purely mechanical in animals, but not in man, is therefore highly implausible.

For Orsini and Bondan[25] stress should be understood as a physiological, neuro-hormonal process through which living beings go through to face an environmental change, in an attempt to adapt to new conditions and thus maintain their homeostasis. Another important classification of stress is related to its nature, form of manifestation and triggered consequences. Thus, stress can be called eustress (when it is a positive event, that is, stress necessary for the survival of the individual facing an adversity) or distress (when the stress triggered is harmful to the organism).

Also according to Professor Sonia, we are all the same, with only the intensity with which we use our faculties to ensure the survival and care of offspring varying from individual to individual and species to species. What differentiates one animal from another is the intensity and frequency of each of these abilities, regulated according to the greater or lesser need to guard against social and environmental hostilities, which, in turn, may also vary, and by analogy with what is known to occur in humans, it is deduced that all living beings endowed with an organized nervous system (diencephalon) can also feel pain. For Peter Harrison, quoted by Sonia Felipe, it is impossible to elaborate a strict theory of pain, and recognizes that the controversy on pain was established from the theory of evolution, which states: the differences between human beings and animals are quantitative and not qualitative; "fish, birds and non-human mammals have their respective centres of pleasure and pain similar to that found in humans. The neurological mechanisms, responsible for pain reactions, are remarkably similar in all vertebrates and some invertebrates".[24]

For Tâmara Levai[26] , "since suffering is an organic state, a psychological alteration or even a sense of unease, the pain implied is not exclusively physical; natural reactions such as crying, screaming or fleeing, may translate a behavior arising from an adverse situation experienced by an animal, human or otherwise. It can be provoked by lack of food, excessive heat or cold, lack of physical exercise, lack of water, frustration, etc. Each of these states is subjectively distinct and each brings its own physiological and behavioural consequence". According to the International for the Study of Pain, cited by Levai "suffering is an unpleasant sensation and emotional experience associated with actual or potential physical damage to the tissues that constitute the organism ".

For Sonia Felipe the mechanistic theses of Descartes, even after three and a half centuries, defended by him, have influenced, until today, the world of experimental science, thus the mechanistic theory of animal nature gives support to the belief widespread among scientists for at least two decades, that animals are devoid of the consciousness of pain and, consequently, can suffer.[24] Almost four centuries after Descartes, neuroscientists and other researchers came together to publish the manifesto "The Cambridge Declaration on Consciousness" proclaiming the existence of consciousness in human and non-human animals. The declaration was publicly announced in Cambridge, UK, on 7 July 2012 at the Francis Crick Memorial Conference on Consciousness in Human and non-Human Animals at Churchill College, University of Cambrigde, by Low, Eldeman and Koch. The statement was published on the Francis Crick Memorial Conference website.[27]

2.1 Oxytocin and the "biology of morality

Long before psychology and modern neuroscience, it was Charles Darwin, with *The Origin of Species*[2] and especially in *The Descent of Man*[29] who brought us the light of reason to put aside our anthropocentric beliefs, reinforced by religious dogmas about our supposed divine superiority, to recognise our true nature and make us more equal to other animals. In Darwin:

> "We have seen that feelings and intuition, the various emotions and faculties, such as love, memory, attention and curiosity, imitation, reason, and so on, of which man prides himself, can
>
> be found in an incipient state, or sometimes in a well-developed condition, in the lower animals."[29]

As much as Darwin could not explain or prove that there was some innate ability to recognize expressions, he recognized the difficulty of communicating expressions among different species, specifically between animals and humans, except in relation to domestic and trained animals, which would have a greater approximation with their owners. According to him: "... as far as I have been able to ascertain, after many attempts, they do not understand any expression that is restricted to the face, with the exception of smiling or laughing, which to some degree they seem to identify, and this determined amount of knowledge was probably acquired, as much by monkeys as by dogs, by the association between harsh and affectionate treatment with our attitudes and such knowledge is certainly not instinctive.[30]

Since the theory of evolution, science has now shed much light on emotions. Neuroscience is particularly concerned with the importance of oxytocin. According to Campos and Graveto[31] , oxytocin is a neuropeptide synthesized in the paraventricular and supraoptic nucleus of the hypothalamus and is released into the circulation through the neurohypophysis, and is also secreted by the central nervous system, functioning as a neuromodulator. According to Dacome and Garcia[32] regions of the limbic system receive innervations from oxytocinergic pathways; these regions are involved in the production of basic emotions in lower and higher animals, such as fear, anxiety, hunger, satiety, pleasure and sexual desire and the peripheral action of oxytocin producing contractions of smooth muscles at the time of delivery and in milk ejection during breastfeeding are commonly described.[32]

For Uvnas-Moberg and Petersson (2005) quoted by Diana Campos and João Graveto[31] (2010); "the oxytonergic nervous system is equally developed in men and women, however, it is much more influenced by female steroid hormones".

Also according to Campos & Graveto[31] , there is a vast body of scientific evidence in humans regarding the effects of oxytocin, "oxytocin modulates social perception, social cognition, social behaviour and, consequently, promotes social closeness and bonding between people" and in addition to anxiolytic effects, "oxytocin modulates social cognitive functions such as trust and emotion recognition". Bratz and Hollander, cited by Campos and Graveto, also add, that oxytocin when administered to children with autism limits repetitive behaviours and improves social information processing, and Domes *et al.* state that the ability to *"read the mind" of* others is a capital foundation for social interactions, and that a single dose of oxytocin is sufficient to cause a substantial increase in the ability to *"read the mind"* and therefore interpret subtle social cues from the eye region of other subjects.[31] The importance of facial and body expressions for communication has also been verified in primates by Frans de Waal.

Known for his research into animal behaviour and consciousness, Frans de Waal in the book *The Age of Empathy: Lessons from Nature for a Kinder Society*[8] , he goes through the history of animal behaviour studies and draws conclusions from his own research with primates stating that animals possess a sense of justice, equality and solidarity, and to the existence of empathy both intraspecific and interspecific, suggesting it is an important evolutionary characteristic for the survival of higher animals, an elaborate observation of what Kropotkin[16] pointed out in the 19th century about mutual support. Which suggests to us that empathy may be one of the drivers for mutual support and altruistic association between individuals.

Through the observation of living animals in their natural habitat, analogously to what was being observed in humans by Charles Darwin *in The Expression of Emotions in Man and Animals*, traces of the expressions of emotions and feelings in animals are sought. According to Darwin: "The ability to communicate between members of the same tribe by means of language was of crucial importance in the development of man. And the expressive movements of the face and body greatly increased the power of language".[30]

Pliny de Toni in *Human Ethology: the example of attachment*, described parental care as one of the main behaviours that helped the genus Homo to survive, increasing social interaction and the development of the species.[33]

2.1 Philosophical contributions of law for moral consideration with non-human animals

In the words of Sonia Felipe[24] , "democracy and justice have been thought of for almost three millennia as an ideal of equal respect that should include all equals, but these are not necessarily all beings capable of suffering harm, pain, suffering and death by the acts of others, alien to their interests, and therefore violating the conditions of their existence, and - according to her - Plato was the first philosopher to recognize that a truly democratic society would bring together men and animals, without discrimination." - Furthermore - "In the 18th century Rousseau claimed that a democratic republic would take the ideal of equality so seriously, that even animals would be respected in their freedom to provide for themselves, and could live together, peacefully, with human beings, who would not be able to harm them." - However according to Felipe - "in the molds of democratic justice, fostered even in the last two or three decades of the second millennium, only human beings were safeguarded the right not to suffer physical exploitation, emotional abuse and untimely death.[24]

For Sonia Felipe, was from Humphry Primatt in his book *"A Dissertation on the Duty of Mercy and Sin of Cruel to Brute Animals"*, 1776, the year of the Declaration of Independence in which Americans proclaim equality and freedom as guiding principles of political order in their country, that the concept of moral rights began to establish legal rights to nonhuman animals. And in 1789, in England, the moral and legal philosopher, Jeremy Bentham writes, *"An Introduction to the Principies of Morals andLegislation"*. In this work, completed but not published by Bentham since 1780, Primatt's central theses are taken up again and in 1834, there was also an edition of the complete text by Humphry

Primatt, which remained ignored by the philosophical academic community until 1892, when Henry Salt wrote *Animal Rights.*[3]

Sonia Felipe[24] identifies three argumentation strategies in the moral status of animals; The conservative follows the moral tradition without questioning it and refuses to make any change in the conception of the place of animals within human morality, conservatives do not recognize that human beings have any duties, either positive or negative, towards animals; The abolitionist position criticises traditional moral philosophy for discriminating against animals of other species, and proposes an end to all forms of animal exploitation, and this position recognises that moral subjects have not only negative duties, those of non-maleficence, but also positive duties, those of beneficence, towards animals endowed with sentience; The welfare or reformist position, in turn, criticizes the traditional forms of animal handling, defending reforms in the capture and confinement system, and in the objectives of experimental research in animal models, as in the proposal of Replacement, Refinement and Reduction, known as the 3Rs[24] However, in Broom & Molento[35] on the definitions of animal welfare, some research to assess welfare cannot take ethical considerations into account in certain processes, because to assess the degree of animal welfare one must first know the consequences of the methods to which these animals are subjected, and only after the results can one state to what extent or what is the best way to use these animals.

Ryder, quoted by Sonia Felipe, considers the use of the animal model in science to be speciesist and clarifies the meaning he gives to the concept he proposes to describe such a procedure:

> "I use the word speciesism to describe the widespread discrimination practised by man against other species, and to draw a parallel with racism. Speciesism and racism are both forms of prejudice based on appearances - if the other individual looks different, they are then considered to be beyond moral parameter. Speciesism and racism (and indeed sexism) ignore or underestimate the similarities between the discriminator and those against whom he discriminates, and both forms of prejudice reveal indifference to the interests of others, and to their suffering."[24,34]

In 1840, the anarchist Proudhon had already thought about the idea of morality in animals in What is *Property, in* Chapter V, § 1, "Of the moral sense in man and animals". Countering religious ideas, he says that the moral nature of man is similar to that of animals, only differentiating us by degree - that is, according to him - the differences between human beings and animals are quantitative and not qualitative, we are only differentiated by our ability to reflect and reason motivated by our justice[17] . However Frans de Waal[8] will state that other animals also make choices and judgements according to their interests.

For Sonia Felipe[24] the ultimate purpose of a free and reasoning nature is to constitute itself in moral nature. According to her, "the only status capable of guaranteeing us a distinction in relation to other living beings - endowed with physical autonomy, but lacking moral autonomy, unable to realize their own vital unity, beyond the determinations of their biological nature - is that of being moral subjects". According to Sonia Felipe, Kant establishes by defining moral autonomy as constitutive of the dignity of beings whose will is liberated by the activity of reason, that is, the activity that establishes ends to be reached through action and this project Kant calls humanism. So, in Felipe's interpretation, nothing that implies the destruction of our will liberated by reason, can be worthy of what we call morality, even less, humanity. Still for her, it must be examined, therefore, whether or not the destruction of animal life can be considered, an act that genuinely represents morality and, therefore, the humanity of a being of moral autonomy. Steven Wise, quoted by her, proposes *practical autonomy* as a criterion for ethical and legal definition of the dividing line that distinguishes living beings, to whom we must recognize and ensure legal rights, from others, to whom we cannot yet conceive of such rights, either because of their mental nature little or nothing is known, either because they are devoid of any form of *practical autonomy*[2,4]

According to Felipe, sensitivity, consciousness, *self* perception, desire and intention constitute some indications or evidence that animals have *practical autonomy* and such indications can be observed through behaviours that result from mental activity, even if in some cases this seems to be minimal. Observation, attention, memory and mental coordination of one's own movement in the natural environment are constitutive abilities of animals capable of making choices, in which Wise recognizes practical autonomy and in relation to which he proposes the constitutional protection of their liberties linked to the enjoyment of this autonomy: non-captivity and the possibility of movement to provide for oneself and for dependents. Self-conscious animals are those capable of knowing that other animals can "see and know". This means that they understand symbols, use a sophisticated language system or something similar, are able to disguise, represent, imitate and solve complete problems[24] . Waal[8] demonstrates through behavioural experiments with primates and other animals this ability to "see and know". For Sonia Felipe conscious animals, which can act and represent significantly, are close to man in the evolutionary scale, have *insight* (think), and these abilities indicate that such animals should be classified in the same scope in which we place human beings with identical abilities [24]

Still according to Sonia Felipe, respecting the practical autonomy or physical freedom of humans and non-humans means preserving: 1) the physical integrity of the subject; 2) the mobility to seek the means of biological subsistence for oneself and one's dependents; and, 3) the conditions necessary for the social interaction of that individual in his or her natural community. Still according to her, for humans, freedom, in the most basic and fundamental sense, means non-slavery, non-imprisonment, non-subtraction of the physical space necessary for subsistence care, mental non-isolation, and social non-isolation, and by violating these limits, one commits the greatest injustice against human beings, since we treat them as slaves and things. And for Wise the same must be applied to the defence of animals as has been established for the defence of humans.[24]

2.2 Some of the recent achievements in animal protection

Democratic constitutions have incorporated relatively well the Universal Declaration of Human Rights of 1984, but are still reluctant to admit in their articles the Universal Declaration of Animal Rights, proclaimed in Brussels, at UNESCO headquarters, on January 27, 1987, and reformulated in April 1989 by animal protection entities around the world, following the initiative of German abolitionists.[24]

Another conquest in the legal sphere was Law n° 9.605/98. The vivisection started to be considered a crime, if not adopted the existing substitute methods. To practice acts of abuse, mistreatment, injury or mutilation of wild, domestic or domesticated, native or exotic animals is punished by detention, from three months to one year, and a fine. The same penalties are applied to whoever performs painful or cruel experiments on live animals, even if performed for didactic or scientific purposes, when alternative resources exist, and the penalty is increased by one-sixth to one-third if the animal dies (Law n° 9.605, 12/02/98 - Environmental Crimes Law - Chapter V, Art. 32°). It can be seen that the environmental legal norm recognizes the cruelty implicit in the experimental activity on animals, counting that there are already alternative techniques to the use of animals in laboratories inside and outside the country.[26,36]

One of the achievements of the reformists (welfareists) in Brazil was the creation of the National Council for the Control of Animal Experimentation (CONCEA) in 2008, which regulates animal experimentation and imposes treatment measures for animal welfare.[37]

CHAPTER 3

Criticism of the anthropocentric model

Peter Singer[38] in his book *Animal Liberation* made some very important reports: to sustain the entire production chain of vivisectionist science there is an industry that shares the interests of banks. According to him, in 1986 the researchers of the Office of Technology Assessment of the American Congress tried to determine the number of animals used in experiments in the USA, and it is believed that the number would be at least 17 to 22 million animals per year, and 22 million animals is what the companies produced annually. The 1988 report by the Department of Agriculture listed 140,471 dogs, 42,271 cats, 51,641 primates, 431,457 guinea pigs, 331,945 hamsters, 459,254 rabbits and 178,249 "wild animals": a total of 1,635,288 animals used in experiments, only 10% of an unclear total, and of the 1.6 million animals declared by the Department of Agriculture as being used in experiments, more than 90,000 are subjected to "incessant pain and stress". And in Japan, a very incomplete study published in 1988 came up with a total of over 8 million animals. Much of the most painful experiments are conducted in the field of psychology; the US National Institute of Mental Health alone has funded 350 animal experiments, spending more than $30 million. And in Britain, where experiments are required to declare the number of "scientific procedures" performed on animals, official government figures show that in 1988, 3.5 million scientific procedures were performed on animals. Another important field of animal experimentation involves the annual poisoning of millions of animals; in Britain in 1988, 588,997 scientific procedures were carried out on animals to test drugs and other products and 281,358 were not related to the testing of medical or veterinary products.

Other authors who have made important reports on the vivisectionist industry were Sérgio Greif and Thales Tréz[39] , and according to sources cited in the book *The True Face of Animal Experimentation,* the National Institutes of Health in the U.S. is the largest funder of animal experiments, they spend $7 billion annually, $5 billion of which is allocated to animal research, And for Brazilian researchers it is difficult to estimate the amount invested in research involving vivisection in Brazil due to the confidential nature of the research, but Greif and Tréz claim that it is one of the areas of greatest funding, perhaps because it is the most expensive for institutions, since, according to sources cited by them, the Social Emergency Fund covered costs of $ 1.7 million in 1995, only with food for animals used in research at federal universities.

For Laerte Levai[36] , specialist in animal rights; "the animal experimentation, defined as all and any practice that uses animals for didactic or research purposes, derives from a methodological error that considers it the only way to obtain scientific knowledge", and according to him, "in Brazil, as it happens almost everywhere in the world, daily thousands of animals lose their lives in cruel experiments, submitted to surgical and toxicological tests, behavioral, neurological, ocular, cutaneous, etc., without any ethical limits or even scientific relevance in such activities". For the researcher Helena Ribeiro[40] , the advances in public health are directly linked to the technological and industrial revolution, and the discoveries that produced considerable progress in human health happened through the study of diseases occurring in populations, case studies, study of cells and bacteria, and the dissection of dead corpses, among other methods. Studies quoted by Singer[14] indicate that in the USA the mortality rate had dropped dramatically among the ten main infectious diseases, with the exception of poliomyelitis, before any new modality of medical treatment had been introduced, concentrating 40% of the drop in mortality rates between 1900 and 1948, with only 3.5% being the result of medical intervention. Singer presumes that this has occurred as a result of improved sanitary conditions and nutrition of the population.[14]

For Peter Singer[14] scientific publications become a favourable source for researchers as they only include experiments considered significant by researchers and editors, because according to Singer, a British government committee found that only about 25% of experiments on animals are published and the Association of the British Pharmaceutical Industry in the midst of debates on the reform of the laws on animal experimentation, through advertising has propagated the idea that human beings today have a longer life expectancy due to the use of animals in experimentation. This information may not be true, because according to Singer and Greif, social and environmental changes, such as improved hygiene and sanitation, and preventive medicine have contributed much more than medical intervention, in the mortality rates of the countries observed.[14,39] Professor Helena Ribeiro in her article Public Health and the Environment presents through the study of history, the importance of environmental quality, the basic notions of hygiene and sanitation, for the quality of life and health of the population.[40] And for Ana Rique[41] *(et al),* cardiovascular diseases are the leading cause of mortality in the world, with studies indicating that a good quality of life (nutrition and physical exercise) can help prevent and control these diseases.

If the success of research came down to its results, human experimentation carried out during World War II in Nazi concentration camps could have made an important contribution to science. For Angélica Rezende[42] and collaborators, after World War II ethical concerns about human beings increased, resulting in the Nuremberg Code and the Declaration of Helsinki and from this the use of animals in research skyrocketed. The Nuremberg Code determined that the results of animal experimentation should be used as the basis for experiments on human beings.

For Thales Tréz and Sergio Greif, another concern regarding animal experimentation is the extrapolation from one species to another, an extremely risky procedure. On the contrary, much of what is tested on animals has harmful effects on them, and not on humans, making it difficult to identify valuable products.[39]

3.1. Methods for the development of science

The first historical documentation of alternatives to the use of animals in the acquisition of knowledge dates back to about 2000 BC, a clay model of a sheep's liver was found in a Babylonian school-temple and was used for teaching divinations. The lobes of the model, the portal fissure, the gallbladder, the cystic duct and part of the hepatic duct are visible. The adoption of the clay model had no animal protectionist purpose, it was merely economical and highly didactic.[39] Today so much is known, that in the USA, 68% of Medical Universities do not use live animals in the disciplines of physiology, pharmacology and surgery.[43]

Currently, in vitro technology is bringing real advances to scientific research and there are several applications of this technology: cancer research, immunology; toxicological tests, where cell viability as well as damage in its structure is used as parameters for analysis of this toxicity. Toxicity testing during development and reproduction can be performed on chicken, fish and amphibian embryos and this methodology has proven to be very important; vaccine production; drug development; study of infectious development; disease diagnosis; study of genetic disorders or diseases. The human placenta can also be used, besides being a source of cells for culture, and material for toxicity and carcinogenicity tests, as a tool for training microsurgical techniques. And the technology for cell culture is being increasingly perfected.[39,44]

The use of the alternative offers advantages such as: greater ease in purifying antibodies; little difference in cost between this and the *in vitro* method, when costs involving maintenance and animal care are considered; when non-protein medium or serum is used a greater production with greater purity in the same period is guaranteed; batch consistency in large scale production.[39]

Bacteria and protozoa are sensitive and mutagenic organisms, allowing them to identify carcinogens. The Ames test, using a Salmonella strain, has confirmed the correlation between mutagenicity and carcinogenicity. Bacteria and protozoa can also be used to estimate vitamin levels in pharmacological and toxicological studies and to identify antibiotics.[39]

Another technique for the production of substances of animal origin uses recombinant DNA technology, which involves the synthesis of protein compounds through genetic manipulation in bacteria. A gene responsible for the production of a certain substance is isolated and inserted into the genetic baggage of these bacteria, which will then produce the substance. Ex: Insulin.[39]

In vitro experiments are appropriate for several studies on the intermediary metabolism used in biochemistry to study the dynamics of enzyme reactions that occur in our biological system. This coupled with mathematical models can contribute to the experimental work by defining variables and testing theories, reducing the cost of these experiments and making them more effective. An example of this is the prediction, through mathematical models, of the structure of proteins, which could predict their physical and chemical properties.[44]

3.1.1 Difficulties and challenges

However, these same substances tested in cells should have their activities studied when applied to a living organism, because *in vivo,* several factors of the organism itself may interfere with the results. Anyway, previous studies *in vitro* help reducing the number of animals used in research.[43] Despite efforts to replace animals in scientific experimentation, many studies still need to be done, especially in the combination of cloning of tissues with recombinant DNA technology, and the question of the response that a tissue or organ independent of an organism can present, in relation to the response that would give a living organism as a whole. The ideal would be to find answers for the biological mechanisms, without the use of live animals, different from the understanding of the bases of the disease of human myasthenia gravis, where there was the involvement of muscles of frogs, synapses of rodents, snake toxin, receptor of electric fish and antibodies of rabbits.[44] However, because this research was independent of each other and did not aim to understand human myasthenia gravis in all cases, it was by pooling the information that the conclusion that human myasthenia gravis is an autoimmune disease was reached.[44] In other words, it was curiosity itself and the wealth of information that led us to the answers at random.

3.2 Inhuman culture and science

Cruelty to animals is worrying, taking into consideration that one of the behaviours that characterise psychopathology in childhood is the cruelty they commit with other children and animals. According to the humanitarian scientist Albert Schweitzer: "Whoever has become accustomed to devaluing any form of life runs the risk of considering that human lives are also unimportant.[45]

For Laerte Levai, the pedagogy of cruelty is consciously or unconsciously inserted in the primer of peoples. From the first acts of gratuitous sadism against insects, to the killing or imprisonment of birds and the ill-treatment of domestic animals, children grow up in a world where violence is part of the urban and rural setting.[36]

For Hilda Morana[46] and collaborators the path to sadism is not clear, although it may be a combination between extreme narcissism and a brain configuration where regions related to empathy are significantly deficient, which would lead the murderer to a total indifference to the suffering of his victims. Perhaps this explains why psychopaths, popularised by Cleckley[47] , are described as very intelligent and rational people, yet they cannot make use of empathy.

The APA (American Psychological Association) classifies antisocial personality disorder (ASPS) as equal to psychopathy and sociopathy.[46] Some characteristics of ASPS are; superficial affect, insensitivity, lack of empathy, lack of remorse or guilt, among others.[46,48]

According to Balenciaga[49] modern society is responsible for the imposition and trivialisation of psychopathy, where profit is above life, creating fragmentation and dispute between individuals for power, contributing to personality disorders. The documentary The *Corporation* by Joel Bakan directed by Mark Achbar and Jennifer Abbott shows a scenario where corporations in a way acquire antisocial characteristics and psychopathy, with a conditioned control so excessive that they directly influence the behaviour and life of society.[50] According to Balenciaga, from a strictly psychiatric point of view there is no pathology, and many psychopaths are neither criminals nor display violent antisocial behaviour as described in criminology textbooks, and the antisocial tendency of psychopathy is present in our culture due to economicist demagogy, which allows antisocial tendencies, violence and psychopathy to go unnoticed and even be seen as normal in a democracy. Besides this, for her, the classification of this disorder applies fundamentally to criminal conduct, leaving aside a variation of emotional and interpersonal problems, and the ideological bias does not allow to identify or classify as psychopaths apparently normal people who are present in fields as varied as politics, economics, medicine and many others, redefining the disorder to an individual and also genetic problem, diluting the social character of such "pathology". Thus, Balenciaga redefines psychopathy as a fundamentally moral and ethical insane madness that targets society through the continued vulnerability of social norms, and psychopaths can be classified simply as bad, rational people who do not have any neurological problems beyond a lack of empathy and callousness, with an absence of concern for others. For

Pinuel, cited by Balenciaga, the emotional world of these individuals is strongly limited, it seems a type of 'social autistic'. In this way, individuals with SPS can also be present in positions of power in society (such as within a company), and can also be adjusted to high social standards, being antisocial towards lower social and/or hierarchical levels, and therefore are not always criminals, as the *'establishment'* itself accepts various forms of exploitation and manipulation of individuals, without remorse and without guilt, preserving SPS in the populations.

According to Lawrence Becker's *social distance* theory, quoted by Sonia Felipe[24] , we would only have moral consideration with members of our own community. Thus Becker justifies the exclusion of animals for being socially distant from the human community, which for Felipe is a contradiction, since we maintain many relationships and interactions with various species, and sometimes even prefer them. Even so, it is important to consider that ethics can represent an "expanding circle" that has as its starting point the moral consideration with its fellows and lastly with other species.

For Gadotti (quoted by Valadão & Milward-de-Andrade)[51] , the "modernisation" of university education ended up displacing the cultural role of universities and neutralised their traditional humanistic orientation, making it subsidiary to the interests of service producers in a world dominated by the industrial mode of production.

For Roxana Valadão, this is the case of the elitist sense of university education, which leads to class behaviour, corroborating the hierarchisation of its structure and highlighting the function of education as a reproducer of values and situations experienced in the stratified society in which it is inserted. This imposition ends up introducing distortions in professional training and leads to a mismatch between the level of training of individuals and the demands imposed by the course of social events.[51]

For Thales Tréz, "the use of animals, as experiments and as a consolidated didactic experience, is increasingly being characterised as a resource and a situation which promotes dehumanisation and alienation", "reinforcing hegemonic postures in favour of maintaining highly questionable conceptions of scientific practice and education".[52] Guided by Bioethics, the research presented below had the following objectives; to ascertain the existence of relationships between the empathy responses and opinions of beginning and final-year biology, pharmacy and veterinary students regarding animal experimentation, and to investigate the existence of relationships between the variations in empathy responses and speciesism. And the hypothesis was that the students at the end of the courses could become even more indifferent to animal suffering, characterizing this Technical-Scientific education, as a speciesist educational process.

CHAPTER 4

The research

The research was conducted between the months of July 2012 and July 2013 at the Universidade Vila Velha in the city of Vila Velha/ES. A questionnaire (Appendix I) adapted from Tréz[53,9] & Nakada[9] was distributed to students of the Biological Sciences, Veterinary Medicine and Pharmacy courses, from the 1st[a] to the 8th period of the university, for them to answer about the use of animals in research and teaching; their preferences and feelings about sacrificed animals. The periods were divided into two categories: Beginning (1st to 4th) and End (5th to 8th).

Question 1 was submitted to descriptive analysis to identify the preference (which animals to replace) of the students. The choices were classified into 4 groups: 'Domestic' (dogs, horses, cats, guinea pigs and rabbits); 'non-domestic' (rats, invertebrates, monkeys, fish, pigeons, pigs and frogs); 'None' (no animals) and 'all' (all animals). In this way speciesism was identified as the choice for one or more specific animals in which the students had the greatest moral considerations, in this case the animals they would prefer to be replaced by alternative methods. The only choice that could characterise an anti-speciesist choice is to replace all animals. Identifying the speciesist choices created 2 speciesist groups, those choosing to replace none and to replace domestic.

To check the students' empathy response, three groups of sensations with certain values (positive with a value of +1; negative with a value of -1; and neutral with values of 0) were available in the questionnaire (Question 2), where only 3 sensations should be marked, building a Likert scale. The negative sensations are: Distress, Guilt, Annoyance, Revolt, Sadness, Difficulty in concentrating. Positive sensations: Admiration, Well-being, Happiness, Pride, Satisfaction and Tranquillity. Neutral feelings: Indifference and Curiosity. With this, the summed score of the students could vary from -3 to +3. This way they were submitted to a t-test using Past 3.0 as follows; empathy x (period and sex); empathy x (course and period and sex); and lastly the relationship between empathy and speciesism also submitted to an ANOVA test; empathy x (none)(domestic)(all).

Questions 3, 4, 6, 6.1 and 7.1 of the questionnaire were subjected to binomial test to analyse the trend of the answers. Questions 5 and 8 were discarded.

4.1 Results and Discussion

Data from 57 projects released by the CEUA-UVV were collected to make the descriptive analysis related to the use of animals in teaching and research at the University Vila Velha, and 281 questionnaires answered by students were tabulated and analyzed, being 106 from biology, 66 from pharmacy, and 109 from veterinary (96 men, 181 women and 4 unidentified).

Course	Projects	Used Animals
Biology	5	674
Pharmacy	9	419
Veterinary Medicine	6	144
Postgraduate Diploma in Animal Science	13	593
Postgraduate Pharmaceutical Sciences	11	536
Postgraduate Diploma in Ecosystem Ecology	13	1984
Grand Total	57	4350

Table 1. Total number of projects and animals used per course.

According to Table 1, the courses that presented the most projects using animals were the postgraduate courses followed by the undergraduate courses. However, the undergraduate courses used more animals than the postgraduate courses in Pharmaceutical Sciences and Animal Science combined.

Most projects are for Theses and Others (internal or outsourced research). And teaching projects were only found in the Veterinary Medicine course (Figure 1).

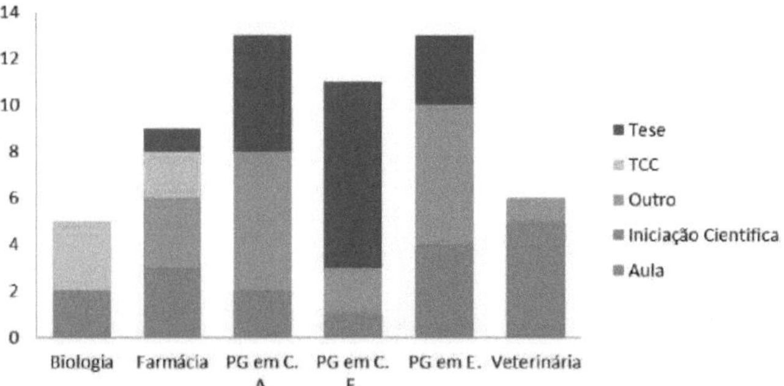

Figure 1. Graph showing the purpose of the animals per course.

During the years 2009 to 2012 a total of 4350 animals were used, and according to Figure 2, 2009 started with only 64 animals and in 2010 there were already 1407. In 2011 a drop to 927 and in 2012 it rose again with 1952 animals used.

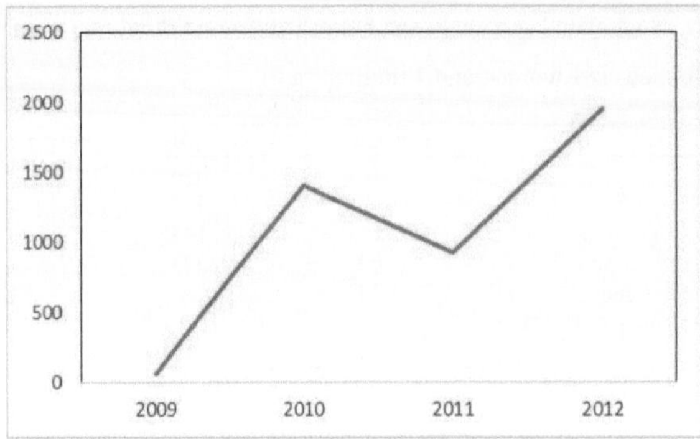

Figure 2. Graph showing the number of animals used each year.

The animals most used in research were classified in this order; mammals (rats, mice, rabbits), followed by fish (tilapia, curimatãs), reptiles (chelonians, lizards, snakes), amphibians (frogs and toads) and lastly domestic animals (horses, cats and dogs). In teaching 91% mammals (rats and mice) were used while 9% were domestic (Figure 3).

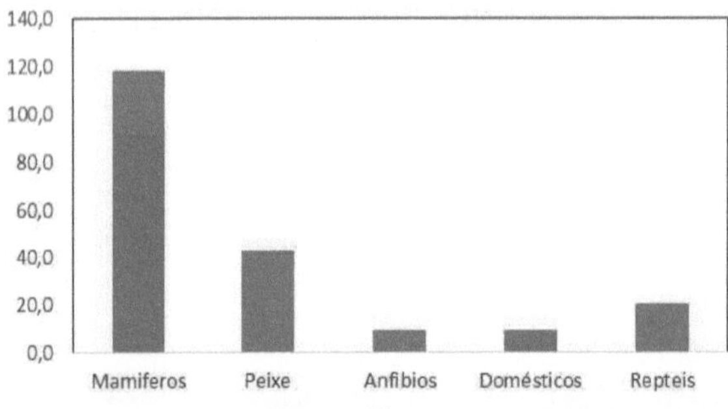

Figure 3. Graph showing the percentage of the types of animals used in teaching and research.

Regarding the origin of the animals, 42.50% were of commercial production origin (breeding), mostly fish intended for toxicological studies. Laboratory animals (rats, mice, rabbits) represent 28% of the animals used in the research, and wild animals (snakes, lizards, chelonians) obtained through scientific collection represent 29.10% of the animals used. And 0.40% were domestic animals (Figure 4). Research involving scientific collection should ideally be replaced in the future by methods of cloning tissues in laboratories, or by simple blood tests, without the need to remove them from their habitat.

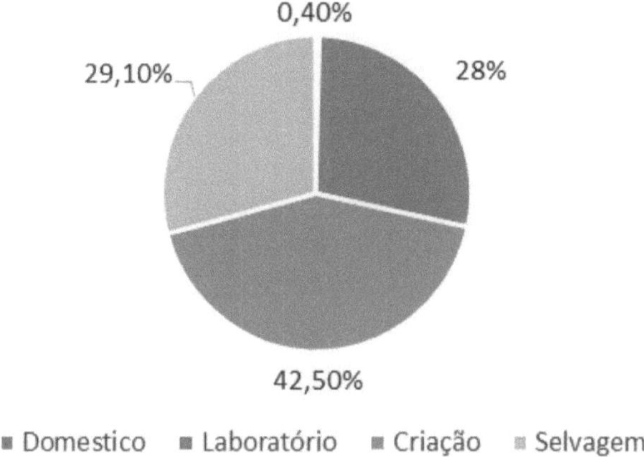

Figure 4. Graph showing the percentage of the origin of the animals used.

As we can see in Figure 6, in 88.5% of the projects the animals were euthanized, while 12% of the projects had alternatives and 11% could not answer. To justify themselves some researchers stated that the laboratories had neither the equipment nor the resources available to perform the alternative methods.

 In 37% of the researches, unknown substances were used, i.e., they were researches that should have considered in *vitro* tests. In 83% of the research involved collection of biological material, where a significant part of the animal was removed (organs and tissues), consequently leading to euthanasia of the animal. In these cases, the substitution by *in vitro* methods for direct tissue evaluation can be done through cloning. In some specific projects, the reproductive organs of chelonians have been removed to identify the sex of individuals, something that can be easily detected by a DNA test or nuclear magnetic resonance.

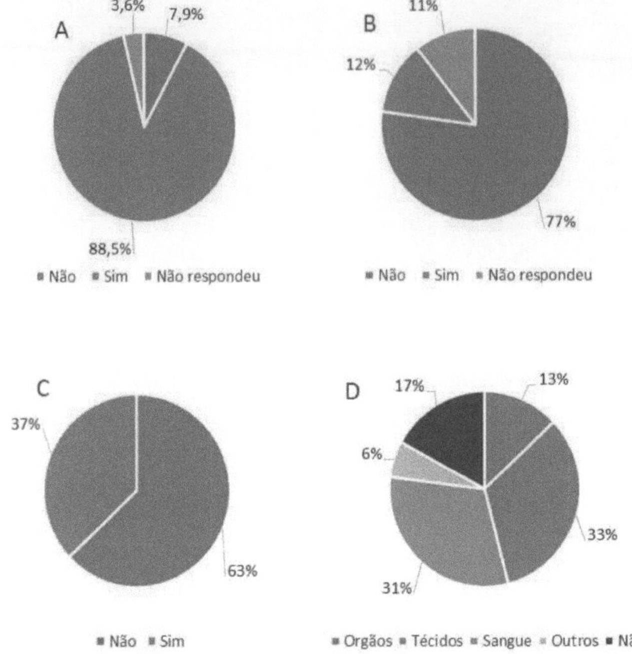

Figure 5. (A) shows the percentage of animals that were euthanized; (B) shows the percentage of studies that had alternatives; (C) shows the percentage of studies that used unknown substances; and (D) shows the percentage of studies that involved collection of biological material.

On animal welfare, 35% of projects involved intentional stress, while only 9% involved intentional pain (Figure 6).

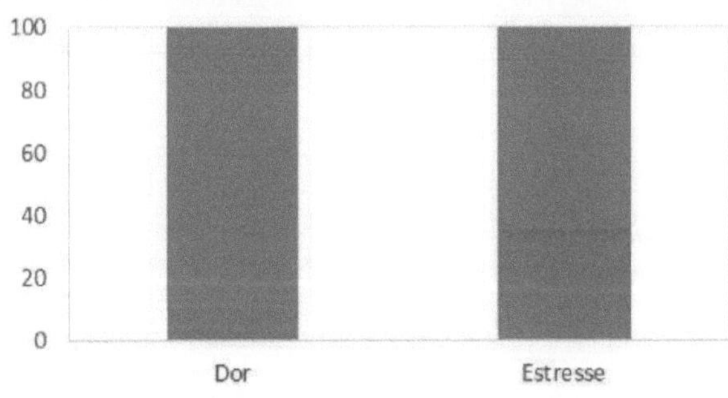

Figure 6. Percentage of experiments that caused intentional pain and stress.

4.1.1. Students' general opinions and knowledge about the use of animals

Regarding the use of animals in research, many students are in favour. However, when there are alternatives, a smaller number would agree with the use of animals, while most of them are unaware of these alternatives (Figure 7).

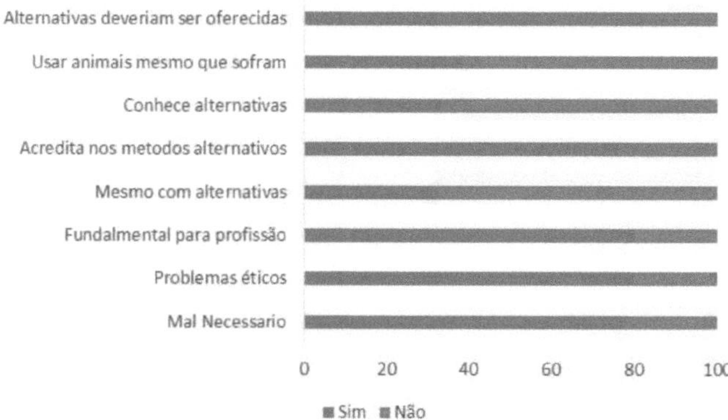

Figure 7. Graph presenting the students' general opinion on the use of animals in research.

There is a difficulty for students to understand what ethics is, as according to 42.5% of students the law is above ethics. This minimizes the chances of a discussion since 63% of students say that teachers affirm the legality of practices. On the other hand, 87.2% of students agree that there is a need to discuss ethics (Figure 8).

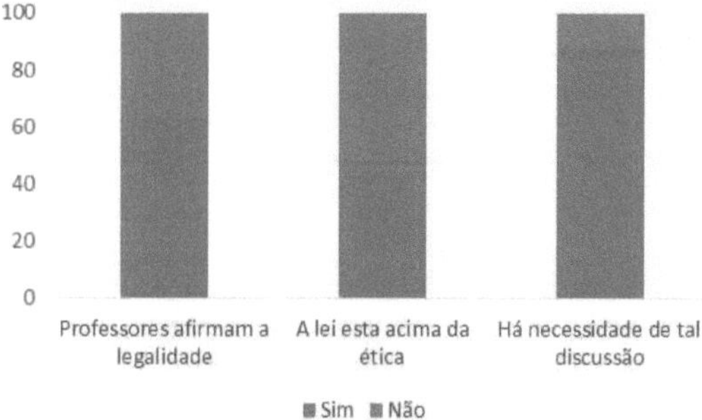

Figure 8. Graph presenting the need for courses in general to discuss laws and ethics.

4.1.2. Identifying speciesism

To identify speciesism in animal experimentation was simple. For 36% of the students the preference is for "Domestic" animals when it comes to replacing them with alternative methods, against 11% for "Non-Domestic". While 26% chose "No Animals", they think the use of all of them is necessary. On the other hand, 22% of students want the replacement of all animals. And lastly "Both" (when the student specifically chose the types of animals and in their count gave a tie, between domestic and non-domestic) with 4% (Figure 9).

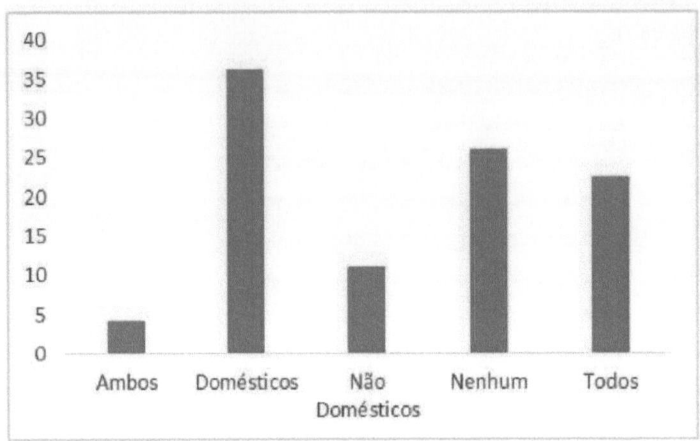

Figure 9. Graph showing the types of animals chosen to be replaced.

If we group the animals into invertebrates, birds, fish, amphibians and mammals, we can clearly see the preference of individuals for mammals (Figure 10), corroborating the research mentioned by Desmond Morris[54] carried out with children, in which they show a clear preference for primates. He also observed that one of the most chosen mammals had been widely advertised by the media, which suggests a strong media/cultural influence. Another observation made by Morris was that the most chosen birds had been penguins, supposedly because of their vertical format capable of resembling our species, and parrots capable of imitating our vocalisation. For Morris these facts may be related to our unconscious desire to anthropomorphise other species. Furthermore, a study by Seyama and Nagayama[55] shows that people can have unpleasant impressions when very realistic human features are observed on computer-generated faces at different levels of reality, and the less strange the set of features the more pleasant the impression that is left. This they call the Uncanny Valley and can be observed in robots, dolls, masks, avatars and virtual reality characters, and anything that can look human.[55] From this, if we ignore respect for diversity (or for what is different) this preference for our "equals" also suggests that it is an adaptive characteristic, as facial and body expressions are more easily identified and read, contributing to better communication.

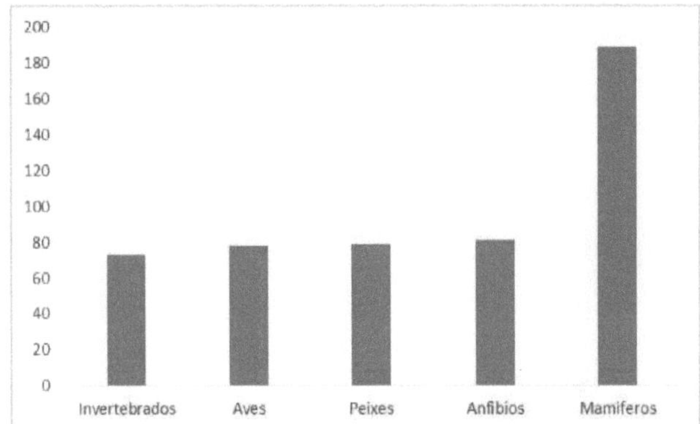

Figure 10. Graph showing total frequency of the groups of animals chosen to be replaced.

Speciesism is also evident when showing the specific choices. Dogs were the most chosen to be replaced, followed by monkeys, guinea pigs and rats (Figure 11). A similar result was presented by Tréz & Nakada.[9]

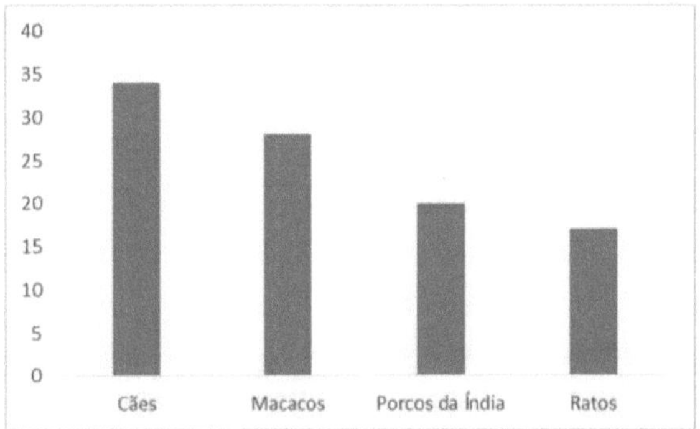

Figure 11. Graph showing percentage of most important animals (among dogs, monkeys, guinea pigs and rats) to be replaced.

The level of speciesism differs between males and females according to the period, as shown in the following graphs (Figure 12). The male gender has a greater acceptance of the use of animals, even if they suffer, and this acceptance increases when compared from the Beginning (47.7%) to the End (56.4%). As for females, there is a low level of acceptance from the beginning (29.9%) to the end (30.8%). Graphically there is not much difference between the sexes, but between the sexes the Binomial test identified a significant difference, men tended to answer "yes" from the beginning to the end and women tended to answer "no" throughout the course.

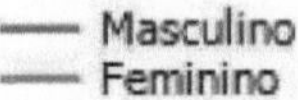

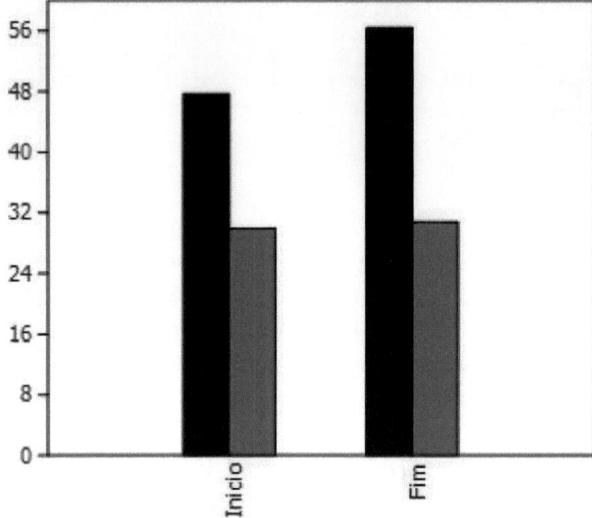

Figure 12. Graphs presents the percentage of overall acceptance of "using animals even if they suffer" for each period in relation to gender.

And also students tended to feel no guilt (16% felt guilty) for the animal's suffering, showing in a way that they did not want to take responsibility for this relationship. On the other hand, a greater number felt distressed (41%) (Figure 13), a feeling linked to unconscious guilt[56] . According to Gaspar[57] , when we consider the relationship between the feeling of guilt and ethics, initially we associate the uneasiness in the human being to the lack of orientation for his/her acting in the world; and from then on we characterize the ethical project as the search for this orientation, and consequently, as an attempt to overcome the uneasiness. For Freud, the feeling of guilt can be represented as the most important problem in the development of civilization.[56,57]

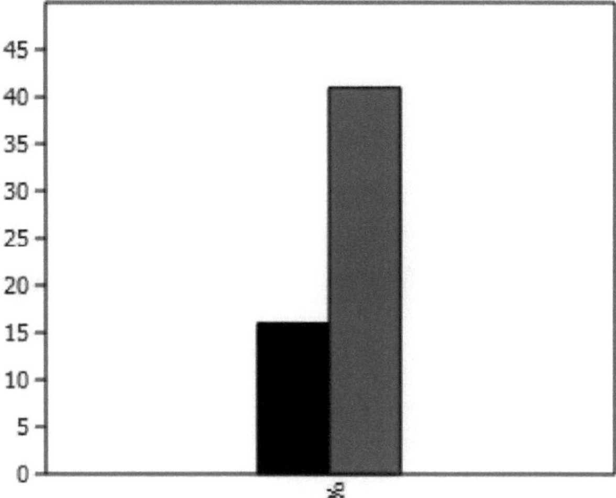

Figure 13. Graph showing the percentage of students who feel guilt and distress over the suffering of the animal.

Overall in relation to the empathy response, the t-test showed significant difference between the sexes at the beginning (p<0.001) and at the end (p<0.001). However it showed no significant difference between the same gender in relation to the beginning and the end, but in comparison, men show significantly higher median (p<0.001) at the End compared to the End of women, who showed no significant variation from the Beginning to the End (Figure 14). This demonstrated that the female gender had a higher resilience compared to the male gender. Resilience being the ability of individuals to prevent, minimize or overcome the harmful effects of adversity, including leaving these situations strengthened or even transformed.[58] Moreover, this variation in response between the sexes may be related to the construction of gender identity[59] , which can influence the characteristics of the person in our society.

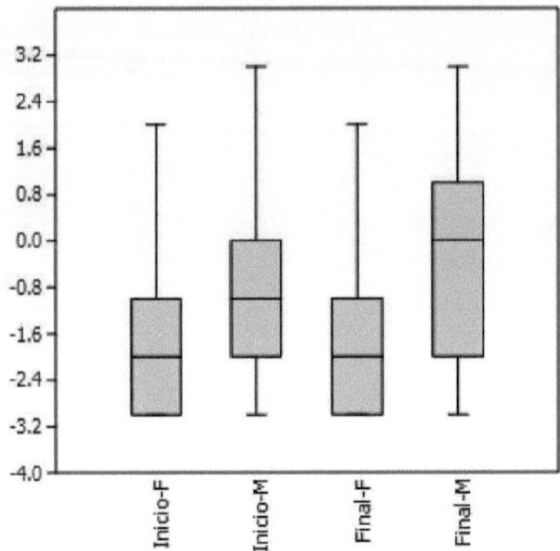

Figure 14. Graph showing the empathy ratio between course periods in general by gender.

Specifically in the veterinary course, the t-test showed a significant difference in the response of empathy between men and women at the beginning (p<0.05), and between men and women at the end (p<0.001), where once again the female sex showed a greater response of empathy. However, between the periods there was no significant difference between the same sex. There was a significant difference between opposite sexes in relation to the period. Between men at the beginning and women at the end, the difference was significant (p<0.05), and between women at the beginning and men at the end (p<0.001), the difference was even greater, due to the fact that the male sex at the beginning showed greater empathy responses (Figure 15).

In the pharmacy course, the t-test showed a significant difference between genders at the beginning (p<0.05), but no significant difference between genders at the end. Between the periods there was no significant difference for the same sex, although the female sex showed greater responses of empathy. There was only significant difference between males at the beginning and females at the end (p<0.05) (Figure 15).

As for the biology course, the t-test showed no significant difference between men and women at the beginning, as well as no significant difference between men and women at the end. It also showed no difference for the same sex in relation to the period nor between the opposite sexes in relation to the period (Figure 15). In general although the medians of the sexes from the beginning to the end are similar, except in the veterinary course for the male sex, there is variation in the levels in all groups.

The male gender at the end of the veterinary course showed a higher level of indifference compared to all the other courses, with some variations off the graph for more and for less in the level of empathy.

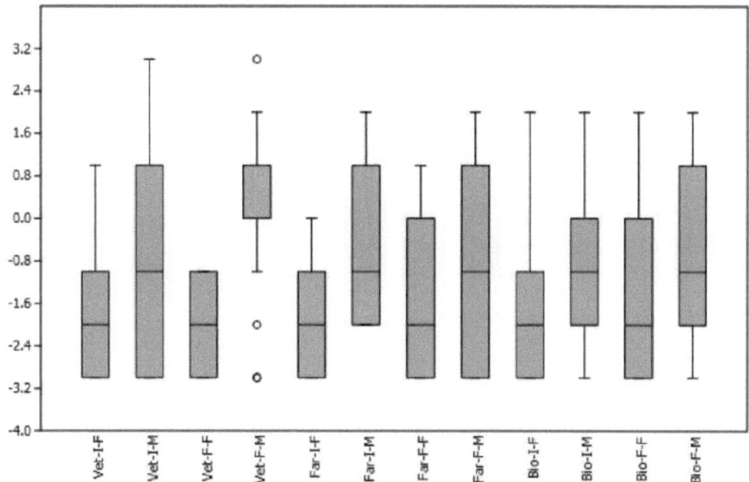

Figure 15. Graph showing the levels of empathy in relation to the courses and period by gender. Vet=Veterinary; Far=Pharmacy; Bio=Biology; I=Start; F=Final; M=Male; F=Female.

Each course and each sex presents its own characteristics, and this relationship with other animals may also be related to other factors, besides scientific education, such as the individuals' own identity and their ideas formed from other social institutions, such as school, family, media, religion, etc.

Thus, considering the influence of the belief system (culture and society) on people's behaviour and emotions, Filippi[60] carried out a study with three distinct behavioural groups (omnivorous, vegetarians and vegans) to evaluate their levels of empathy towards human and non-human animal suffering. The result showed that the Vegetarian (less speciesist) and Vegan (anti-speciesist) groups presented higher empathy coefficients (Figure 10).

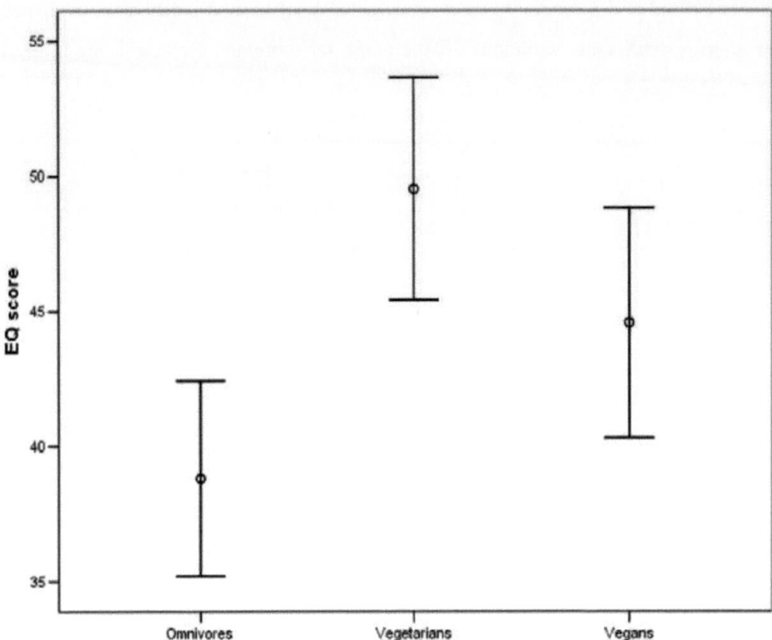

Figure 10. Graph showing the empathy coefficient between distinct behavioral groups: Onivores (speciesists), Vegetarians (less speciesists) and Vegans (non-speciesists) (FILIPPI et al, 2010).

This result can be compared with ours (Figure 11), as considering the response "All" as being anti-speciesist, it was the one that demonstrated the highest response of empathy towards animal suffering. To identify the level of empathy towards speciesism, the ANOVA test showed a significant result (p<0.001) among students who chose to replace all animals (anti-speciesist), students who chose preferably domestic animals (speciesist) and students who chose no animals (more speciesist). Being that the significant difference by t-test between "All" and "Domestic" (p<0.05) is less compared to "All" and "None" (p<0.001). There is higher concentration of empathy responses for those who chose "All" and "Households", however those who chose "Households" show higher values for indifference. While those who chose "None" were well dispersed with little empathy, tending towards greater indifference (Figure 11).

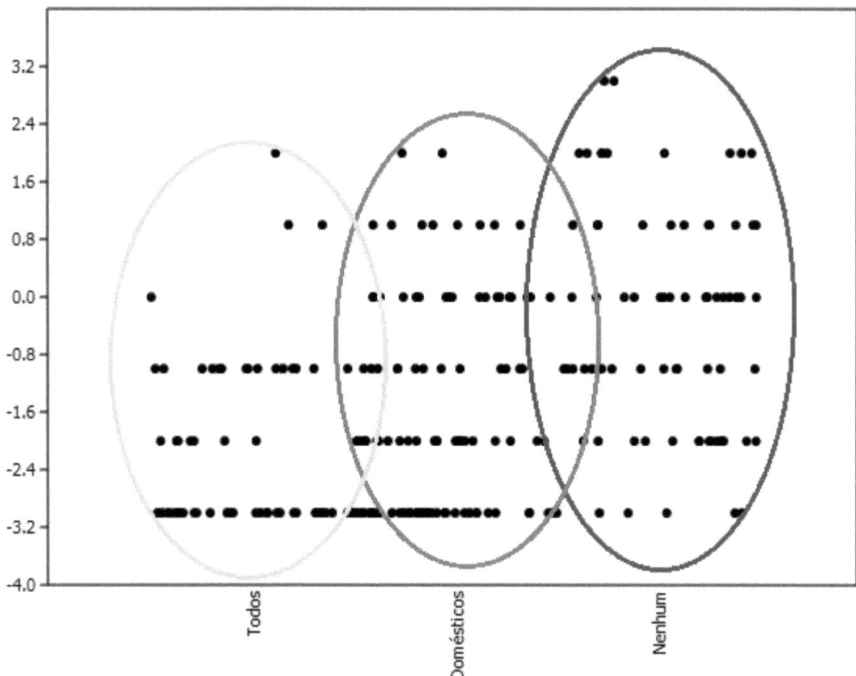

Figure 11. Graph presenting the level of empathy of students who chose between replacing 'All animals', 'No animals' and preferably the 'Domestic' ones. With negative numbers presenting greater empathy as they represent negative feelings when asked about the sacrificed animals.

CHAPTER 5

Conclusions

Despite the high acceptance of animals as models for scientific experiments, under the allegation of a necessary evil, when the issue is suffering, opinions in favour of the use already decrease. The high acceptance of the use of animals in divergence with the low knowledge on animal welfare, ethics and alternatives, takes away from the student the possibility to have an opinion based on information and to develop a critical awareness on what they do, making it impossible to make ethical decisions. It was seen in some cases that the difference between periods showed variation in the students' empathy response, especially in relation to gender. The study showed that gender influenced significantly more than periods in moral consideration towards other animals. Which may indicate to us that the technical-scientific education that these students go through, may influence their moral considerations towards other animals, but this is not the only variable.

We have seen that each course and each sex presents its own characteristics, and their moral considerations with other animals may also be related to other factors, other variables besides scientific education, such as the individuals' own identity and their ideas formed from other social institutions, such as school, family, media, religion, etc.

The fact that the female gender shows higher values of empathy and greater resilience in this research does not mean that the female gender is biologically more adapted to this behaviour. We can consider that it is the differentiated education to which the two sexes are subjected in our society.

Considering then people's belief systems, we can verify (assuming that everyone is speciesist) that less speciesist individuals (with some moral consideration towards other animals without distinction of species) may present better empathic responses or be more sensitive to the suffering of another animal. If this is true, we can suggest that empathy towards a particular group is a behaviour that can be developed during one's life, according to one's individual experiences (e.g. a speciesist upbringing, just as a racist or sexist upbringing). Thus, even though the different forms (and species) of life may have a greater difficulty in empathic communication between them, either because of the difficulty of recognizing the foreign body and its manifestations, or because they are seen as resources, our knowledge, ideas and values transform our perception due to alterations in our sensations, which may interfere in our choices and consequently in the change of our behaviour.

If we consider that we have an interspecific social relationship with other animals, in speciesist practices we can identify some characteristics of TPAS, such as lack of remorse or guilt, superficial affection, insensitivity and lack of empathy. If we consider the intelligent and rational use to violate these animals for our own benefit, intraspecific TPAS in our relationships becomes even more evident. The greatest concern for the trivialization of this behaviour is exactly the difficulty of identifying individuals who really suffer from Intraspecific Trauma Syndrome from those who do not, but keep the same practices and repress their moral considerations, becoming hostages of an activity imposed by the *'establishment'*.

Further studies are needed, mainly in the development of a methodology/technology to better understand the functioning of our cognitive aptitudes for empathy and perception. And the study of biology and ethics-oriented psychology can still lead us to seek significant transformations in human actions. Meanwhile, as the study of behaviour in the field of biology and psychology is not enough to solve the problem of violence, the philosophical and legal reinforcement of Biodiret. Only in this way will we be able to develop a human society more and more in balance with nature, and in this specific case, to develop a more humane science.

CHAPTER 6

References

1 CAPRA, Fritjof, **The Turning Point**. Editora Cultrix: 1982.

2 REGO, Teresa C. **Vygotsky: a historical-cultural perspective of education**. 16 ed. Petrópolis, RJ, 1995-2004.

3 VIGOTSKY, Lev Semenovich. **The social formation of mind: the development of higher psychological processes** / organized by Michael Cole [et all]. 7ª ed. São Paulo: Martins Fontes, 2007.

4 KOHLER, Wolfgang. **Wolfgang Kohler: Psychology** / organized by Arno Engelmann. Grandes cientistas sociais n. 4. São Paulo: Ática. 1978.

5 COSTA, D. S.; GAMA, J. D.; SOUZA, L. C. A.; ALMEIDA, R. M.; DINIZ, R. T. B.; SOUZA, R. C. **Ética, Moral e Bioética.** *Jus Navigandi,* Teresina, year 2, n. 21, nov. 1997.

6 CHAUÍ, Marilena. **Invitation to Philosophy.** Ática, São Paulo, 2000.

7 FELIPE, Sonia T. **Por uma questão de princípios: alcance e limites da ética de Peter Singer em defesa dos animais.** Florianópolis: Boiteux, 2003.

8 DE WAAL, Frans. **The Age of Empathy: Lessons from Nature for a kinder society.** Companhia das Letras, 2009.

9 TRÉZ, Thales; NAKADA, Juliana. Perceptions about animal experimentation as an indicator of the anthropocentric-speciesist paradigm among teachers and students of Biological Sciences at UNIFAL-MG **Rev Alexandria.** 2008; 1(3): 3-28.

10 LIMA, Kênio E. C.; MAYER, Margareth; CARNEIRO-LEÃO, Ana M.; VASCONCELOS, Simão D. Conflito ou convergência? Perceptions of teachers and graduates on ethics in the use of animals in zoology teaching. **Investigations in Science Teaching**. V.13. n.3. pp.353-369.   2008. Available at:

<http://www.if.ufrgs.br/ienci/artigos/Artigo_ID200/v13_n3_a2008.pdf>. Accessed on: 5 Dec 2013.

11   BALCOMBE, J. **The use of animals in higher education**: problems, alternatives &
recommendations. Washington: The Humane Society Press, 2000. Available at:
<http://www.humanesociety.org/assets/pdfs/parents_educators/the_use_of_animals_in_hig
her_ed.pdf>.

12   DINIZ, Renata et al. Animals in practical classes: can we replace them with the same
quality of education? **Rev. bras. educ. med.**, Rio de Janeiro, v. 30, n. 2, 2006. Available at
                              :<http://www. scielo.br/sci elo.php?script=sci_arttext&pid=S0100-
55022006000200005&lng=en&nrm=iso>.

13   SIQUEIRA-BATISTA R, Rôças G; Gomes AP, Cotta RMM, Messeder JC; Mattos EA. Are
environmental bioethics and deep ecology paradigms for thinking the 21st century?
Ensino, Saúde e Ambiente. 2009,2(2):44-51.

14   SINGER, Peter. **Ethical Life.** Ediouro Publicações S.A. 2002.

15   LEAL, Pablo Campos. **Anarchy, Ecology and Veganism: Contributions of Élisée Reclus for
a Bioethical Vision of Space**. International Colloquium: Élisée Reclus and the Geography of the
New World. 2011. Available at:
<http://redebrasilis.net/MemoriasReclusSP2011/campos.pdf>. Accessed on: 7 Jan. 2014.

16   KROPOTKIN, Piotr. **Mutual Support: a fact of evolution**. Porto Alegre, São Sebastião:
Editora Deriva, A Senhora Editora, 2012.

17   PROUDHON, Pierre-Joseph. **What is property?** 1ed. Martins Fontes, 1988.

18   ALVES, Paola Biasoli. The ecology of human development: natural and planned experiments.
Psicol. Reflex. Crit., Porto Alegre, v. 10, n. 2, 1997. Available from
<http://www.scielo.br/scielo.php?script=sci_arttext&pid=S0102-
79721997000200013&lng=en&nrm=iso>. Accessed on: 29 Oct. 2013.

19   POTTER, Van Rensselaer. **Bioethics: bridge to the future**. São Paulo: Edições Loyola, 2016.

20   MATOS, José Claudio Morelli. **Instinct and reason in human nature, according to Hume
and Darwin.** Sci. stud., São Paulo, v. 5, n. 3, Sept. 2007 . Available at:
<http://www.scielo.br/scielo.php?script=sci_arttext&pid=S1678-

31662007000300002&lng=en&nrm=iso>. Accessed on: 29 Sep. 2012.

21 PEGORARO, O. A. **Ética e Bioética**. Editora Vozes. 2002.

22 DINIZ, D. **Bioética: Fascínio e Repulsa**. Acta Bioética, year VIII. n.1. 2002.

23 NAPOLI, Ricardo Bins Di. Animals as persons? The place of animals in the moral community. **Princípios. Journal of Philosophy.** v.20 n.33 Jan/Jun 2013.

24 FELIPE, Sonia T. **Ética e experimentação animal: fundamentos abolicionistas.** Florianópolis: Ed. Da UFSC, 2007.

25 ORSINI, Heloísa; BONDAN, Eduardo Fernandes. Fisiopatologia do estresse em animais selvagens em captivo e suas implicações no comportamento e bem-estar animal - revisão da literatura. **Revista do Instituto de Ciências da Saúde.** 2006. 1(24): 7-13.

26 LEVAI; Tâmara Bauab. **Vítimas da Ciência: Limites éticos da experimentação animal.** 2ª edição. São Paulo. Editora Mantiqueira, 2006.

27 LOW, Philip. 2012. **The Cambridge Declaration on Consciousness**. Available at: <http://fcmconference.org/img/CambridgeDeclarationOnConsciousness.pdf>. Accessed on: 24 Sep 2012.

28 DARWIN, C. R. **On the origin of species by means of natural selection, or preservation of favored races in the struggle for life.** London: Murray, 1859.

29 DARWIN, Charles. **The Descent of Man.** London, 1871.

30 DARWIN, Charles. **The Expression of the emotions in man and animals.** Trad. By Leon de Souza Lobo Garcia. São Paulo: Companhia das Letras, 2000.

31 CAMPOS, Diana Catarina Ferreira de; GRAVETO, João Manuel Garcia do Nascimento. Oxytocin and human behavior. **Revista de Enfermagem Referência.** v. 3, n. 1, p.125-130, Jul. 2010.

32 DACOME, Ocimar Aparecido; GARCIA, Rosângela Fernandes. Modulatory Effect of Oxytocin on Pleasure. **Revista Saúde e Pesquisa**, v.1, n.2, p.193-200, May/Aug. 2008.

33 DE TONI, Plinio Marco et al. Human ethology: the example of attachment. **Psico-USF (Impr.)**, Itatiba, v.9 ,n.1, Jun.2004. Available at: <http://www.scielo.br/scielo.php?script=sci_arttext&pid=S1413-82712004000100012&lng=en&nrm=iso>. Accessed: 15 Oct. 2012.

34 FELIPE, Sônia T. Fundamentação ética dos direitos animais. The legacy of Humphry Primatt. *Revista Brasileira de Direito Animal.* v.1, n.1, jun/dez 2006. Available at: <http://www.animallaw.info/journals/jo_pdf/Brazilvol1.pdf>. Accessed on: 28 Oct 2012.

35 BROOM, D.M.; MOLENTO, C.F.M. Animal welfare: concept and related issues - Review. Archives of Veterinary Science v.9, n.2, p.1-11, 2004.

36 LEVAI; Laerte Fernando. **Direito dos Animais**. 2ª edição. São Paulo: Editora Mantiqueira de Ciencia e Arte. 2004.

37 CONCEA. **Conselho Nacional de Controle de Experimentação Animal.** 2008. Available at: <http://www.mct.gov.br/index.php/content/view/310553.html>.

38 SINGER, Peter. **Animal Liberation.** Trad. Marly Winckler. Porto Alegre, São Paulo: Lugano, 2004.

39 GREIF, Sergio; TRÉZ, Thales. **The true face of animal experimentation.** Rio de Janeiro: Educational Society "Fala Bicho", Sept. 2000.

40 RIBEIRO, Helena. Public health and environment: evolution of knowledge and practice, some ethical aspects. **Saude soc.**, São Paulo, v. 13, n. 1, Abr. 2004. Available at :<http://www. scielo.br/scielo.php? script=sci_arttext&pid=S0104-12902004000100008&lng=en&nrm=iso>.

41 RIQUE, Ana Beatriz Ribeiro; SOARES, Eliane de Abreu; MEIRELLES, Claudia de Mello. Nutrição e exercício na prevenção e controle das doenças cardiovasculares. **Rev Bras Med Esporte**, Niterói, v. 8, n. 6, Dec. 2002. Available from <http://www.scielo.br/scielo.php?script=sci_arttext&pid=S1517-86922002000600006&lng=en&nrm=iso>.

42 REZENDE, Angélica Heringer de; PELUZIO, Maria do Carmo Gouveia;

SABARENSE, Céphora Maria. Animal experimentation: ethics and Brazilian legislation. **Rev. Nutr.**, Campinas, v. 21, n. 2,   Apr.   2008. Available at: <http://www.scielo.br/scielo.php?script=sci_arttext&pid=S1415-52732008000200010&lng=en&nrm=iso>. Accessed on: 28 Sep. 2012.

43   MAGALHÃES, M.; ORTÊNCIO FILHO, H. **Alternativas ao uso de animais como recurso didático**. Arq. Ciênc. Vet. Zool. Unipar, Umuarama, v. 9, n. 2, p. 147-154, 2006

44   MORALES, Marcelo M. Alternative methods to the use of animals in scientific research: myth or reality? **Cienc. Cult. [online].** v. 60, n. 2, pp. 33-36. 2008.

45   MARTA, Taís Nader; MAZZONI, Henata Mariana de Oliveira. Serial killers: a legal or psychological issue? **Revista USCS - Direito.** a. 10 n.17. jul./dez. 2009.

46   MORANA, Hilda C P; STONE, Michael H; ABDALLA-FILHO, Elias. Disorders of personality, psychopathy and serial killers. **Rev. bras. Psiquiatr.**, São Paulo, 2006. Available at:<http://www.scielo.br/scielo.php?script=sci_arttext&pid=S1516-44462006000600005&lng=en&nrm=iso>. Accessed on: 22 Oct. 2012.

47   CLECKLEY, H. **The mask oh sanity**. St. Louis, MO: Mosby, 1988.

48   GOMES, Cema Cardona; ALMEIDA, Rosa Maria Martins de. Psychopathy in men and women. **Arq. bras. psicol.**, Rio de Janeiro, v. 62, n. 1, abr. 2010. Available at: <http://pepsic.bvsalud.org/scielo.php?script=sci_arttext&pid=S1809-52672010000100003&lng=en&nrm=iso>. Accessed on: 23 Oct. 2012.

49   BALENCIAGA, Inmaculada Jáuregui. Psicopatía: Pandemia de la Modernidad. **Nómadas: revista crítica de ciencias sociales y jurídicas.** Madrid, n19, 2008. Available at: <http://www.ucm.es/info/nomadas/19/ijbalenciaga.pdf>. Accessed on: 23 Oct. 2012.

50   NASCIMENTO, Luiz Felipe. Psychopathic Company *versus* Citizen Company. **Rev. Gestão Social and Environmental.** v1,   n1, pg. 19-29, abr. 2007. Available at: <http://www.revistargsa.org/rgsa/article/view/13/3>. Accessed on: 23 Oct. 2012.

51   VALADAO, Roxana; MILWARD-DE-ANDRADE, Roberto. The teaching of Biology: its relations with animal experimentation and environmental advocacy. **Cad. Saúde Pública**, Rio de

Janeiro, v. 6, n. 4, Dec. 1990. Available at:
<http://www.scielo.br/scielo.php?script=sci_arttext&pid=S0102-311X1990000400006&lng=en&nrm=iso>. Accessed on: 29 Sep. 2012.

52 TRÉZ, Thales de Astrogildo e. Experimentando a desumanização: Paulo
Freire e o uso didático de animais. **R. B. E. C. T.**, vol 4, núm 2, mai./ago. 2011. Available at:
<http://revistas.utfpr.edu.br/pg/index.php/rbect/article/view/585/698>. Accessed on: 29 Sep. 2012.

53 TRÉZ, Thales. **O Uso de Animais Vertebrados como Recurso Didático na UFSC:
Panoramas, Alternativas e a Educação Ética.** Course Conclusion Paper - Universidade Federal
de Santa Catarina, Santa Catarina, 2000.

54 MORRIS, Desmond. **The Naked Monkey**. Circle of Books S.A. 1967

55 SEYAMA, Jun'ichiro; NAGAYAMA, Ruth S. The Uncanny Valley: Effect of Realism on the
Impression of Artificial Human Faces. **Presence**, vol. 16, n. 4, aug. 2007.

56 GELLIS, André; HAMUD, Maria Isabel Lima. Feeling of guilt in Freudian work:

universal and unconscious. **Psicol. USP**, São Paulo, v. 22, n. 3, Sept. 2011. Available at:
<http://www.scielo.br/scielo.php?script=sci_arttext&pid=S0103-65642011000300011&lng=en&nrm=iso>. Accessed on: 27 Oct. 2013.

57 GASPAR, Taís Ribeiro. The feeling of guilt and ethics in psychoanalysis. **Psyche (São Paulo)**,
São Paulo, v. 11, n. 20, jun. 2007. Available at:
<http://pepsic.bvsalud.org/scielo.php?script=sci_arttext&pid=S1415-11382007000100004&lng=en&nrm=iso>. Accessed on: 27 Oct. 2013.

58 ANGST, Rosana. PSYCHOLOGY AND RESILIENCE: A literature review. **Psicol. Argum.**,
Curitiba, v. 27, n. 58, p. 253-260, jul./set. 2009.

59 LOURO, Guacira Lopes. Gender and sexuality: contemporary pedagogies. **ProPosições**,
Campinas, v.19,   n.2, Aug. 2008. Available at:
<http://www.scielo.br/scielo.php?script=sci_arttext&pid=S0103-73072008000200003&lng=en&nrm=iso>. Accessed on: 16 Oct. 2013.

60 FILIPPI M, Riccitelli G; Falini A, Di Salle F, Vuilleumier P, et al. The Brain Functional Networks Associated to Human and Animal Suffering Differ among Omnivores, Vegetarians and Vegans. **PLoS ONE** 5(5), 2010.

UNIVERSIDADE VILA VELHA
THE USE OF ANIMALS AS A DIDACTIC AND SCIENTIFIC
RESOURCE

STUDENT SURVEY

Course: Area you intend to follow:
Period: ___________
Gender:( )F( )M

1. If you have any objections to the use of animals in teaching and/or research. Please tick the species(s) you consider most important to be replaced.

( ) Dogs(	) Invertebrates	( ) Pigs
( ) Mice / rats	(worms, insects, etc.)	( ) Guinea pigs
( ) Horses( )	Monkeys	( ) Frogs / toads
( ) Rabbits( )	Fish	( ) All animals
	( ) Cats	( ) Pigeonsshould be replaced

2. What sensations do you experience when coming into contact with animals that will be sacrificed or have been sacrificed in laboratories? Choose three of them:

( ) Admiration( )	Curiosity( )	Indifference( )	Satisfaction
( )	Difficulty( ) Uncomfortable( )		Tranquil
( )	Concentration	( ) Pride( )	Sadness
( ) Guilt( )	Happiness	( ) Revolt	

3. About the use of animals **in research**. Answer (Y Yes; N No; NS Don't Know):

Y N NS

Is it a "necessary evil"?

Are there ethical problems with the use of animals?

It is fundamental to your profession

Should animals continue to be used even if alternatives can be applied?

Do you believe in the viability of alternative methods to the use of animals?

Do you know of alternatives to the use of animals?

Do you agree with the use of animals, even if they have to suffer from the procedures?

Should alternatives be offered to students opposed to the use of animals?

4. About the use of animals **in teaching**. Answer (Y Yes; N No; NS Don't know):

Y N NS

Is it a "necessary evil"?

Are there ethical problems with the use of animals?

It is fundamental to your profession

Should animals continue to be used even if alternatives can be applied?

Do you believe in the viability of alternative methods to the use of animals?

Do you know of alternatives to the use of animals?

50

Do you agree with the use of animals, even if they have to suffer from the procedures?

Should alternatives be offered to students opposed to the use of animals?

5.What would **primarily** prevent you from questioning the use of animals in the classroom?

( )   Unawareness of alternatives ( ) Does not see reasons to discuss, because it does not

( ) Fear of reprimand by teachers has a   problem with the use of animals

( ) Afraid of the opinion of classmates   ( ) Other:

( ) Does not feel entitled to criticise the methodology of the teacher

6. Do teachers usually affirm the legality of practices with animals? ( ) Yes ( ) No

6.1 Do you agree that the law is above ethics? ( ) Yes ( ) No

Do teachers usually address ethical issues before, during or after animal experiments?

( ) Always ( ) Rarely ( ) Never

7.1 Do you believe there is a need for such a discussion? ( ) Yes ( ) No

8.How often do you see animal rights campaigns on social media?

( ) Always ( ) Rarely ( ) Never

Printed by Books on Demand GmbH, Norderstedt / Germany